Freshwater Ecology

Freshwater Ecology

A Scientific Introduction

Gerry Closs
Department of Zoology
University of Otago
Dunedin
New Zealand

Barbara Downes
School of Anthropology, Geography
and Environmental Studies
University of Melbourne
Melbourne
Australia

Andrew Boulton
Ecosystem Management
University of New England
Armidale
Australia

Blackwell
Publishing

© 2004 by Blackwell Science Ltd
a Blackwell Publishing company

BLACKWELL PUBLISHING
350 Main Street, Malden, MA 02148-5020, USA
108 Cowley Road, Oxford OX4 1JF, UK
550 Swanston Street, Carlton, Victoria 3053, Australia

First published 2004 by Blackwell Science Ltd
Reprinted 2004

Library of Congress Cataloging-in-Publication Data

Closs, Gerry.
    Freshwater ecology : a scientific introduction / Gerry Closs, Barbara Downes,
Andrew Boulton.
        p. cm.
    Includes bibliographical references (p.).
  ISBN 0–632–05266–X (pbk. : alk. paper)
    1. Freshwater ecology. I. Downes, Barbara J., 1958– II. Boulton, Andrew J.
III. Title.
    QH541.5.F7C66 2003
    577.6—dc21

                                                                    2003002574

A catalogue record for this title is available from the British Library.

Set in 10 on 12.5 pt Palatino
by Kolam Information Services Pvt. Ltd, Pondicherry, India
Printed and bound in the United Kingdom
by MPG Books Ltd, Bodmin, Cornwall

The publisher's policy is to use permanent paper from mills that operate a sustainable
forestry policy, and which has been manufactured from pulp processed using acid-free and
elementary chlorine-free practices. Furthermore, the publisher ensures that the text paper
and cover board used have met acceptable environmental accreditation standards.

For further information on
Blackwell Publishing, visit our website:
http://www.blackwellpublishing.com

Brief contents

Full contents

Contents

Contents

Contents

Preface

Ecological scientists require two broad types of skills and knowledge – the more general skills and knowledge related to the practice of doing science, and specialist skills and knowledge related to the particular question they are investigating. General scientific skills relevant to ecology include an understanding of the scientific method, an appreciation of how the scale of sampling influences our perception of ecological patterns, an ability to collect data in a consistent and rigorous manner, and the skills to analyse, interpret and communicate results effectively. Over the course of a scientific career, an individual researcher may work on a wide variety of research questions, and hence be required to develop a variety of specialist skills and knowledge specific to each research project. However, the general scientific skills will be used over and over again, constantly being refined over a scientist's career.

Development of a student's general scientific skills can be something of a haphazard process. The pressure on teachers to impart large amounts of specialist knowledge to students in a relatively short time can leave only limited opportunities to develop the more general skills. Often, we either assume that the students will acquire such skills along the way as they complete various class projects, or read through material that is assigned to them. However, unless students are deliberately made aware of the need to develop general scientific skills, then they are unlikely to strive consciously to develop them.

A major motivation in writing this text is our belief that the development of general scientific skills underpins a successful scientific education. Students need to understand how we do science so they can seek to improve their general scientific skills from one project to the next. In doing so, they become adept at applying and adapting their science to the different situations they are likely to encounter over their careers. A key feature of this book is an initial emphasis on the scientific method and the effects of scaling on the measurement of ecological patterns in a freshwater context. Such skills and appreciation of scale effects are essential for any ecological scientist irrespective of their specific field of research.

The use of these general skills is then emphasized throughout the rest of the book with a focus on the approaches used to explore various research ecological questions. In this introductory text, we have deliberately minimized the amount of specialized information provided to maintain a focus on the intellectual approaches used to tackle different scientific problems and to avoid overloading students with detail. Each of our chapters might form the basis for a lecture, with reading lists provided so that students can then read the primary literature. In addition, other more advanced freshwater ecological and limnological texts that present specialized information for advanced students are readily available.

Numerous people have contributed indirectly and directly to the development of this book. Some deserve special mention. Our undergraduate and postgraduate students have played a key role over the years in the writing of this book, constantly pushing us to keep up to date, and reflect upon and refine our teaching and research methods. Several people read and made comments on various chapters, and we particularly wish to thank Alex Flecker, Nancy Grimm, David Hart, Ben Ludgate, Chris Peterson, Craig Stow, Colin Townsend, Mike Winterbourn and two anonymous referees. At Blackwell Science, Ian Sherman, Sarah Shannon, Katrina McCallum, Rosie Hayden and Jane Andrew played key roles at various stages of the project. Our respective university departments – Zoology, University of Otago; Anthropology, Geography and Environmental Studies, University of Melbourne; and Ecosystem Management, University of New England – allowed us the time and flexibility to undertake and complete this project.

Finally, a special thank you to Robyn Sperling and Debra Panizzon for their patience and support throughout.

Gerry Closs
Barbara Downes
Andrew Boulton

Part 1 The Tools of Freshwater Ecological Science

Freshwater ecological scientists use a variety of tools to do their work. These include intellectual tools such as scientific methods of thinking and presentation of arguments and an understanding of the effects of scale. Other practical tools include knowledge of the properties of water and the types of habitats and organisms likely to be present in the system under study. The learning required to master the use of these scientific tools continues throughout a scientist's life. Often, the types of tools and their sophistication also evolve over time as the technology and knowledge base of our science expands and develops. Indeed, one of the most personally rewarding aspects of a career in science can be the great satisfaction that comes with using the various scientific tools effectively in novel and challenging situations.

In Chapters 1–5, we introduce some of the 'tools' that are regularly used in freshwater ecology. In Chapter 1, we discuss the logic and philosophy of the scientific method. There is no universally agreed process on how we should do science, but an awareness of the ways in which a scientific argument can be developed is crucial for developing an understanding of the strengths and limitations of both your own work, and the work of others. In Chapter 2, we address the issue of spatial and temporal scales of study, and how this can influence our choice of sampling sites, period of sampling, sampling equipment and even the types of organisms we study. Developing survey programs that sample communities at spatial and temporal scales relevant to the questions that we are asking is one of the most important skills an ecologist needs to develop. Errors in design of these studies or an inappropriate choice of scale can lead to erroneous conclusions and hamper our understanding seriously.

In Chapters 3–5, we describe some of the basic features of the freshwater systems in which we work. The initial step in any scientific study is observation. Observation provides us with the basic information that we use to develop models and hypotheses. However, to observe effectively, a basic knowledge of the system under study is required. We need to know what are likely to be the primary habitats within an aquatic system, what are some of the physical and chemical characteristics of those habitats, and some of the likely characteristics of the organisms that live there. Armed with the approaches from the first two chapters and the knowledge from the next three, we will be ready to explore some of the themes and questions that puzzle freshwater ecologists today.

Chapter 1 *What is ecological science?*

1.1 Introduction

Numerous books on science, especially introductory ones, contain many 'facts' but do not address why this information is interesting, relevant or useful. Science, however, is not about the simple accumulation of facts, and to present facts in the absence of any framework for why they are important is to starve the description of science of its most exciting elements. It also impedes the beginning student from developing an ability to start thinking critically about ideas.

In this book, we present the scientist as detective and science as an activity designed to answer questions and to solve problems, not as one in which the goal is mass accumulation (and ultimate memorization) of information. Some memorization of information must occur, but facts ought to be presented in the context of the theories upon which they were designed to shed light. Consequently, we begin with a definition of science and a discussion of scientific methods before reviewing how this is practised in the science of ecology. We focus on the kinds of questions asked by ecological scientists and the sorts of methods they use.

Before we begin, we must emphasize two things. First, this chapter cannot do justice to all of these topics – the material is only introductory. Serious students of science should expect to read steadily in scientific methodology throughout their career and to consider many more issues than we can devote space to here. Second, if you were to take 100 ecologists and ask them to describe the main methods upon which their science is based, you would never get universal agreement. It is important to realize that the descriptions of ecological science we present are not some universally agreed approach used without variation everywhere in the world. Our views are influenced by our biases in both background and in reading, but we are at least revealing them! We do believe though that many practising ecologists would accept the approach we outline below as a fair description of our science.

1.2   What is science?

There is no universal or fixed definition of science on which everyone agrees, but there are attributes that distinguish science from other areas of intellectual enquiry. First, science concerns itself with natural, not supernatural, phenomena. If we invoke some omniscient being to explain what is happening, then we have stepped outside the boundaries of science. Second, predictions arising from explanations or hypotheses should be testable – that is, there must be some relatively objective method (we discuss this below) by which we can collect observations that will test our ideas. Third, many scientists insist on repeatability as a basic attribute. Different researchers should be able to repeat an experiment and get the same result.

In practice, the latter two of these requirements cause difficulties. It is possible to suggest theories, for example, that are impossible to test because the technology to do so does not exist. Some ideas in particle physics remain untested until someone dreams up a way to do the definitive experiment. Repeatability is a basic criterion for sciences like physics and chemistry, where experimental conditions and experimental subjects or materials can be exactly replicated at different times and places. It is a much more difficult criterion to satisfy in environmental sciences like ecology. Environmental scientists deal with experimental subjects like animals, plants and environmental locations that are all inherently variable, which makes it much more difficult (although not impossible) for these scientists to repeat an experiment and get the same result.

Before discussing 'the scientific method', we should comment on the distinction often raised between 'basic' and 'applied' science. Traditionally, basic science is seen as concerned with questions arising purely from curiosity. Whether and how findings may be applied to solve problems for human beings is not a central concern. Applied science concerns itself directly with the latter sorts of problems, and has sometimes been viewed as a 'lower class' of science. We suggest, as have many other scientists before us, that distinguishing between these two sorts of questions is pointless. Scientific advances in understanding come from all quarters – from individuals preoccupied with grand questions as well as those trying to solve what seemed, at the time, a small technological problem. Additionally, chance, accidental discovery and history all play roles in scientific discovery, and may happen to anyone with an open mind.

In ecology, 'applied' questions are those where we ask how we can detect and repair or modify the impacts of human beings on the natural environment. We believe such questions should be viewed simply as ecological questions, and they should be subject to the same requirements of rigorous testing.

1.3 Scientific methods

The first thing almost any book that discusses scientific methods will say is that there is no one scientific method. While this is true, there is evidence that some methods have been better than others in advancing scientific progress as measured by improved understanding or successful prediction. Below we describe briefly the forms of reasoning used in science, and their association with scientific methods. However, we emphasize again that this material is introductory.

1.3.1 Forms of reasoning: induction and deduction

We practise **induction** when we collect a large number of careful observations, discern patterns from those observations and then infer general laws or theories from those patterns (Fig. 1.1). We argue from our particular observations to general statements about the world. In its simplest form (Chalmers (1999) calls it 'naïve' induction), the principle of induction can be stated thus:

> If a large number of As have been observed under a wide variety of conditions, and if all those As without exception possess the property B, then all As have the property B. (Chalmers 1999)

Hence, induction uses the frequency with which some observation has been made in the past to draw inferences about what we shall see in the future. We rely on our experience under a wide array of conditions and the collection of unbiased facts. Induction does not use formal logic. We

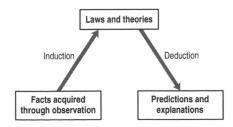

Fig. 1.1 A visual representation of inductive and deductive reasoning in science. (After Chalmers 1999.)

are saying that because we have observed something many times before that we shall observe it again tomorrow.

Simple induction relies on scientists having the proper training to make observations that are objective and that this allows them to draw general laws from 'the facts' in ways not open to those without scientific training. Few scientists would practise or defend simple induction, and induction as a scientific method involves somewhat more complicated practices than we have described here. However, there are several problems with induction.

First, we know that observations are affected markedly by past experience and expectations of the observer. Two observers of the same event can 'see' quite different things, and the human senses are not always reliable instruments. This means that even careful training does not guarantee we will not introduce bias into our observations. Induction provides no context for experiments, one of the main ways in which scientists can find things out. Finally, and perhaps of most concern, induction does not separate the context of the discovery of a theory from the justification for that theory. In the words of Peter Medawar, a famous English biologist who wrote much about the practice of science, induction means 'That which leads us to form an opinion is also that which justifies our holding the opinion' (Medawar 1982, p. 87). However, the process of thought in induction that supposedly leads us to 'truth' just as often can lead us to error, and induction provides no way to help us recognize when our ideas are wrong. Medawar commented that four fifths of his time in science was wasted on ideas that were incorrect (Medawar 1982). Scientists who test their ideas – who provide, if you like, nature with an opportunity to show us we are wrong – often have the same experience. We are wrong far more often than we are right, and when we are right we are usually only partially right. The observation that we are frequently wrong sits uneasily with a scientific method that does not signal us clearly when we are on the wrong track. It is also one of the major reasons why those who attack science, either because there are major disagreements between scientists or because some theory has shown to be incorrect, fundamentally misunderstand what science is all about.

Deduction argues from the general to the specific (Fig. 1.1). We use general laws or theories to make predictions about what we will see in our specific observations. Unlike induction, deduction makes use of a piece of formal logic (Box 1.1). Also unlike induction, we deduce the falsity of our ideas from our observations rather than 'the truth'. To use Chalmer's (1999) example, if our theory is that all ravens are black and we observe a raven that is not black, then our observation has disproved our theory and we have made a logically valid deduction. We can discard that hypothesis. This deductive logic is used in the scientific method known as **falsificationism**. In falsificationism, we try to disprove ideas

---

Box 1.1

The logical basis of disproof

Disproof is based upon a piece of logic called *Modus tollendo tollens*:

> If some statement *p* implies (or causes) some other statement *q*, the discovery that *q* is false allows us to reject or negate *p*.

In falsification, we try to disprove the null hypothesis using the same logic:

From our null hypothesis (*p*), we make a prediction about what we expect to find in our data (*q*). If we collect observations that contradict these expected data (i.e. *q* is false), then we can reject our null hypothesis (we negate *p*).

(formally stated as null hypotheses – we shall talk more about these below) – that is, we test them and we learn by being wrong. Note that an hypothesis that is not disproved is not proven, because it is logically impossible to prove any but the most trivial of statements (think of what would be required to prove every raven in the world is black!). Instead, an hypothesis that survives testing is retained as a viable idea up until it does not survive a test we place upon it. We proceed then, not by building a case to prove an idea, but by disproving alternatives. A minimum criterion for hypotheses, then, is that they must be potentially falsifiable – it must be at least possible to collect observations that are inconsistent with them if they are wrong.

The use of formal logic – rather than our experiences – as the basis for drawing inferences is the greatest strength of falsificationism. Falsificationism is the method we use and shall focus upon in this book, but falsificationism has its problems. Many scientific hypotheses are far more complicated than simple statements like 'all ravens are black'. Complex hypotheses often include a main claim and several subsidiary claims. All of the latter may need to be correct for the main claim to be correct. If we falsify our hypothesis, it is possible that one of the subsidiary statements was incorrect rather than the main claim itself. If we have not tested the subsidiary statements independently, then it may be unclear whether one or more of those is incorrect or the main claim is incorrect. In some cases, 'blaming' subsidiary statements for being wrong has been a *post hoc* way for scientists to evade falsifying a favoured idea. This is clearly a problem.

1.3.2 'Strong inference'

Above, we have argued that falsificationism is a good basis for scientific method because it is based on logically valid inferences, rather than our experiences. This view was emphasized by Platt (1964), who argued that some sciences were making rapid progress, as measured by their pace of discovery. These sciences were distinguished by two main features, which collectively produced what he called **strong inference**. The first of these features was the use of experiments that could provide unambiguous

---

Box 1.2

Experiments and unambiguous results

In well-designed experiments, all possible outcomes each have only one very likely interpretation or explanation. We gain the strongest degree of inference from such experiments. In all other ways of collecting data, such as surveys or quasi-experiments (in which, for example, experimental treatments cannot be assigned randomly to subjects), there will be more than one explanation (possibly many) for any differences that are observed. This ambiguity provides weaker inference, as we will not know which possible explanation is the correct one.

---

results (Box 1.2) – that is, each experiment clearly has the potential to falsify a hypothesis if it is wrong. The second important feature of these sciences was that practitioners had multiple working hypotheses, not just one. This meant that falsification of one hypothesis left researchers with clear alternative explanations to investigate. We can view scientific investigation then as a logical tree. At the start, we may have many hypotheses to explain some pattern or phenomenon we have observed – there are many branches on the tree. We proceed to conduct a series of clean experiments that eliminates one hypothesis after another – we are, in effect, pruning the branches of our tree. Coupled with falsification, the use of multiple hypotheses provides very strong inference. Another desirable outcome is that if there is no single working hypothesis, then there is less chance that an individual hypothesis becomes of such singular importance that its rejection becomes a blow to someone's ego. The incentive to try to make the facts fit a favoured theory becomes reduced, and science becomes much more the exciting detective story that it actually is.

1.3.3 Where do hypotheses come from?

Above, we commented that there should be a clear demarcation between the way in which we make inferences – test hypotheses – and any 'method' we use to think up the hypotheses in the first place. Where do ideas come from? There is no established method that guarantees we will think of a new idea, but there are certain activities that are associated with creative thinking. For example, scientists who read broadly outside of their own immediate literature (such as literature from other sciences) are more likely to make an intuitive leap. Scientists who read literature in the humanities, play music, write poetry or go to the opera may also be exposed to other ways of thinking that will provide analogies for problems in their own field. Chance events or accidental discovery ('serendipity') can play strong roles in suggesting new ideas. The separation of the creative process from the rigour of tests means we can see how imagination and play can coexist with strict scientific methods without automatically compromising objectivity. We can still demand any new ideas survive the rigours of falsificationism.

1.3.4 Biases in science

A common perception of science is that proper training guarantees that scientists are objective and unbiased, but there is plenty of evidence that scientists, as a group, are not unbiased. Culture, race, gender, political system and socioeconomic background all influence the conclusions that scientists, as a group, may reach. The problem seems particularly evident in areas of research with important biological or social implications, like intelligence or sexual behaviour. Hubbard (1990) describes how older prevailing views about gender roles and mate selection in animals – which focused almost exclusively on active competition between males and viewed females as virtually passive receptacles for sperm – were very likely the result of biology being dominated by almost exclusively male viewpoints. Gould (1981) describes how some white, Anglo-Saxon, Western male scientists managed to 'find out' that white, Anglo-Saxon, Western males had higher intelligence quotients than anybody else.

Such episodes have seen science and scientific methods attacked as 'no better' than any other areas of study, including religious beliefs like voodoo. The acknowledgement of bias, however, is not a fatal blow to science. We should remember that any knowledge – even 'facts' we currently regard as 'truth' – is always at least partly wrong. Well-established ideas will be eventually partly or wholly replaced by others that can explain more of the evidence. Rejecting science completely because some scientific research might be biased is unreasonable.

Nevertheless, the repudiation of such an extreme viewpoint is not an excuse for scientists to espouse the alternative, extreme viewpoint that 'proper' science is completely objective and without bias. There seems little doubt that the practice of science would be greatly enhanced if more of its members were formally educated about the sorts of biases they might be bringing to their field. Almost certainly, the best outcome would be if scientists came from every sector of society because the biased (self-interested) conclusions of one group are then more likely to be challenged by another.

1.4 The nature of the science of ecology

1.4.1 What is ecology?

Ecology is the scientific study of the distribution and abundance of living organisms and their relations with their environments (Box 1.3). Ecologists may study populations of organisms (defined as a group of interbreeding individuals that are all members of the same species), groups of species living together in a particular habitat (assemblages or communities) or

---

Box 1.3

The bastardization of the word 'ecology'

'Ecology' was coined by the German zoologist Ernst Haeckel in 1866 and means literally 'in house'. It was used to describe the growing interest in understanding the relations between flora and fauna and their natural environment. Unfortunately, 'ecology' has since come to mean a lot of other things. In the popular media and in some of the humanities and social sciences, 'ecology' and 'ecological' are used to describe types of environmental ethics, green political parties or recycling schemes. Of course, words have always been pilfered from one intellectual area to others – that is the evolution of language. Nevertheless the association of the word 'ecology' with 'green' environmental ideologies in the minds of most of the public is unfortunate. There are many public, and sometimes emotive, battles between conservation groups and pro-development groups. Ecologists (and the scientific societies that represent them) can contribute what is known about the likely effects of human impacts on the environment. However, it is difficult to do this and have even a chance of being viewed as independent, if 'ecologist' is thought to mean 'greenie' or any person having a strong conservation ethic.

---

whole ecosystems (the complete array of species and the environment in which they live). Ecologists ask questions such as:

■ What regulates population sizes?
■ Why are some species common and others rare?
■ Why are species more abundant in some places than others?
■ How do interactions among species occur?
■ Why do some environments contain many more species than others?
■ How are food chains built?
■ How are nutrients and energy moved from one part of an ecosystem to another?

Ecology overlaps in content and interest with several other fields including zoology, botany, behaviour and evolution as well as other biological sciences such as physiology, genetics and molecular biology. Ecology also shares many interests with other environmental sciences like biogeography, geomorphology, oceanography and limnology, the latter being the scientific study of inland aquatic environments.

1.4.2   Research traditions in ecology

Before we discuss the ways in which ecologists have tackled these questions, it is worth examining briefly the historical roots of ecology. Research traditions develop within particular fields and are handed on from teacher to student. One of the interesting things about ecology is that its research traditions come from a very wide range of different areas. For example, population ecology has a research tradition based upon the use of mathematical equations and 'bottle experiments', laboratory experiments looking at population growth in relatively simple organisms (like the single-celled Protozoa). On the other hand, ecosystem ecology sprang from field-based studies, especially those looking at the composition of

---

Box 1.4

Who and what are ecologists?

'Ecologist' is not a profession with which the general community is particularly familiar, especially given the uses to which the term 'ecology' is put these days (see Box 1.3). Nevertheless, in most developed nations, ecologists require university-level training in ecology, biology (botany and/or zoology) and disciplines like chemistry, geography, mathematics and statistics and experimental and survey design. Depending on where you live, this may require a 3- or 4-year undergraduate Bachelor of Science degree followed by one or more postgraduate research degrees (such as a 1-year Honours degree or 2-year Master of Science degree) where the student conducts an ecological research project. Research ecologists usually require a

research-based Doctor of Philosophy (usually a further 3–4 years).

There are many professional organizations that represent ecologists (the Ecological Society of America, the Ecological Society of Australia, the British Ecological Society). Many ecologists also belong to professional societies devoted to scientific study of particular environments (e.g. Australian Society for Limnology, American Society of Limnology and Oceanography) and some also belong to organizations looking at improved environmental management (Society for Conservation Biologists, Society for Ecological Restoration). Some of these societies also practise professional certification. Almost all of the societies have Internet web pages and most produce peer-reviewed journals that report ecological research.

---

forests or the dynamics of lakes. Related disciplines like oceanography and limnology were also making detailed field observations about fauna and flora but were less concerned with interactions among species than with the effects of the physical environment. As a result, limnologists, for example, were (and still are) much more likely to talk to aquatic chemists and hydrologists than they are to ecologists working in different environments.

This varied set of research traditions means that ecology encompasses a rich and varied array of research questions and approaches. However, it also means a bewildering array of relevant journals, the numbers of which have exploded in the last few decades. Habitat-based reading is a common problem in ecology – for example, marine ecologists do not read the freshwater literature and vice versa. The result is that theories may arise in one area of ecology without recognition or use in another area of ecology. The plethora of journals and lack of citation across disciplines can be quite daunting to the beginning student trying to make sense of the field. There is no easy solution to this problem except to read as broadly as possible across a wide array of different journals and books, and to be aware that one's training may mean exposure to only a subset of particular ideas and information and viewpoints – those that your teachers think important!

1.5   Types of studies in ecology

It is appropriate at this point to examine briefly the sorts of approaches and skills that ecologists use.

1.5.1 Natural history

Natural history is where we take detailed, careful observations of species in the field. Ecologists with a 'good eye' are those able to see species in nature and make observations that others miss. It is a useful skill and one that is hard to learn. A good ecologist always ventures out with a notebook to record chance observations. Chance observations may lead to new ideas or a new viewpoint.

1.5.2 Empirical surveys

Chance or casual observation may sometimes mislead us, so we must also be able to carry out proper surveys to see if patterns we think we have observed are real. The design of such surveys can be very complex, especially if we wish to look at several different patterns simultaneously. Consequently, an empirical ecologist must be well versed in sampling theory and survey design and the statistical analysis of such data (Box 1.5). Some questions cannot be answered with manipulative experiments, and so cleverly designed surveys may be the only avenue by which we can test ideas. Experiments are impossible, for example, with many ecosystem level questions or those that relate to processes acting over large scales because it is usually impossible to manipulate entire ecosystems or the large-scale factors involved.

1.5.3 Modelling

A whole branch of ecology uses mathematics and/or computer simulations to try to model ecological systems. Such modelling may be as simple as a single linear equation or may involve a large array of equations linked together. Such models usually have to include many simplifying assumptions about individual species or the environment under consideration. The simplicity of the assumptions, coupled with the density of

---

Box 1.5

Why do we need sampling theory?

In ecology, we usually wish to find out various things about populations of organisms, which collectively often number millions of individuals. We cannot possibly examine every single individual within a population, but sampling theory shows us that we do not need to do this anyway. Sampling theory allows us to take a sample from a defined group of things called a statistical population (which could be an actual population of a species, or a group of locations, or anything else we are interested in) and draw inferences from that sample about the rest of the population that we have not directly examined. Sampling theory is thus an extraordinarily useful and powerful tool. Understanding how to take samples whilst abiding by the assumptions of sampling theory is an essential skill for ecologists.

equations and complexity of the mathematics or computing involved, has meant that many empirical ecologists completely ignore these models. Nevertheless, mathematical models can tell us whether certain propositions are or are not correct given a set of particular starting conditions – this sort of proof cannot be gained from simple verbal statements. Additionally, some mathematical models and computer simulations can produce quite novel ideas. Mathematical modelling works well when tailored to a specific ecosystem and accompanied by field testing of both assumptions and predictions, but direct collaborations between mathematical modellers and field ecologists are still relatively rare.

1.5.4 Experiments

An experiment involves the deliberate manipulation of one or more factors to discover the effect this has on some variable of interest. In well-designed experiments, we can draw very strong inferences about the effects of particular factors on our subjects (Box 1.2), and experiments have a big role in falsification (see Section 1.3.1). Experimental design refers to a plan for assigning subjects to experimental conditions and the statistical analysis associated with that plan (Kirk 1995). It is a very important skill to learn, as we shall discuss below.

1.6 Scientific method in ecology

1.6.1 The falsificationist approach in ecology

We describe here in more detail how the falsificationist approach can be used in ecology. Our description borrows heavily from a thoughtful paper written by Underwood (1990) because this is a particularly clear explanation of the process. However, similar philosophies have been expressed in other more recent publications (e.g. Peters 1991). We use Underwood's flow chart (Fig. 1.2) to illustrate each of the steps outlined below.

The procedure begins with **observations** – data that demonstrate a pattern or problem that we wish to explain. For example, we may find that a species occurs more often in one sort of habitat than another. We may find that one habitat has many more species of predators than another. These observations are patterns that deserve explanation. The observations should be reliable – that is, based upon data that clearly demonstrate the pattern is real and not a product of biased observation (Box 1.6).

In the next step, we think of as many different explanations, theories or **models** for our observations as we can. It is important to think of many explanations because of the multiple working hypotheses component of

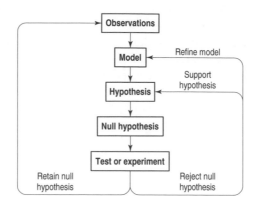

Fig. 1.2 Steps in falsification. (After Underwood 1990.)

strong inference (see Section 1.3.2). Models can take many forms – they may be mathematical equations (as described in Section 1.5.3) or verbal statements, and they may be very simple or quite complex. For example, take our observation above that individuals of a species (let us call it X) occur more often in one habitat (habitat A) than another (habitat B). Models to explain this pattern could include: (i) species X needs sunlight to survive and there is more sunlight in habitat A than B; (ii) a predatory species lives predominantly in habitat B and eats species X; (iii) there is more food for species X in habitat A than in habitat B so that X has higher survival rates in habitat A; and so on. We may gain these models from reading the literature, from our own knowledge of the species involved or from brainwaves. This is the imaginative part of science, where we get to be creative. Experienced ecologists can often, if pushed, think of at least a dozen different explanations for a pattern they observe.

Box 1.6

When a pattern isn't a pattern

One of the hardest things to learn, as a beginning student, is that casual observation does not mean that patterns we think we see actually exist. We might observe a few individuals eating a particular food but not realize that they are unrepresentative of the rest of the population. We may happen to observe many individuals in one location without realizing that this is simply a chance aggregation or that we simply did not see the species in other places. Because we do not demand strong evidence in everyday life ('seeing is believing'), it is hard to reject the notion that simple observation equates with 'fact'. However, once human beings have decided they can 'see' a pattern in nature, they are more likely to see (and remember) observations that are consistent with it and forget (or not see) those that are not.

The only solution is to collect observations in a systematic way that avoids such bias, so that if the pattern does not really exist, data inconsistent with our expectations are as (or more) likely to be collected as data that are. Learning how to carry out such unbiased data collection is a very important part of an ecologist's – indeed, any scientist's – training.

14

The next step is to generate one or more **hypotheses** from each model. Each hypothesis is a **prediction** of what we expect to see in some new circumstances if the model concerned is correct. To continue with the examples above, we can make the following predictions for model (i): if model (i) is correct, then if we shade some areas of habitat A (so that light levels are comparable to those in habitat B) then we predict that numbers of species X in those areas will decline to match those seen in habitat B. We could make similar predictions for the other two models, involving the manipulation of predators or of food levels. Hypotheses should contain predictions that are **unique** to a particular model – if more than one model makes the same prediction, then we have no way of discerning which of them is correct if that particular prediction is found to be true.

To understand the next step, we need to revisit the section on deduction above (see Section 1.3.1). We cannot prove our hypotheses but we can try to falsify them using the logic of disproof. To do this, we generate a **null hypothesis** for each hypothesis. The null hypothesis is one of no difference or no change. For example, the null hypothesis for the hypothesis given above would be that there will be no difference in the numbers of species X in shaded vs. unshaded areas of habitat A. If we carry out our **test** and show that this null hypothesis is incorrect (i.e. we do see more individuals of species X in unshaded vs. shaded areas) then we have provided some support for our hypothesis and our model that it is sunlight that is responsible for the pattern we have observed. We can now refine our model to make it more precise (for example, we might carry out the experiment at other locations to see if we get the same result). If we do not reject the null hypothesis – we see no difference between shaded and unshaded areas – then we must reject the hypothesis and the model from which it was derived. We have discarded one of our possible explanations.

Tests, which may be manipulative experiments or cleverly designed surveys, are critically important. Usually we will need to make use of inferential statistics to decide whether or not to reject the null hypothesis. Additionally, designing experiments that will test a prediction cleanly without introducing artefacts (unintended consequences of some treatments) can be very difficult indeed. For example, we may wish to exclude a species of predator by using cages to keep it out in some places, but cages (especially in flowing waters) create altered flow conditions, cause sediment to be deposited within the cage and may act as a substrate for algal growth that might change the food supply, among many other things. All of these changes might explain any differences we note between caged and uncaged areas besides the absence of the predator. We will need to use various **controls** that will allow us to isolate the effect of interest from all these artefacts. Where manipulative experiments are impossible and we must rely on surveys instead, considerable ingenuity is required to design a survey that will provide relatively unambiguous results. For these reasons, a good grasp of statistics, the sampling theory upon which they are based, the manner by which we use test statistics to reject or retain hypotheses, and experimental and survey design are all essential skills for ecologists.

1.6.2 Induction in ecology

Above, we indicated that we will focus on falsification in this book, but we are not suggesting that induction therefore has no role or is not practised in ecology. Some ecologists collect data in ways that are much more like induction – that is, they take a large number of careful observations and try to infer general theories from those observations. No attempt is made to falsify a previously stated hypothesis. It may be that ecology benefits from a mix of the two approaches. Theories suggested by an inductive approach can be tested under new circumstances – that is, we can use a blend of induction and falsification in ecology. For the beginning student, it is important to understand the difference between the approaches and to recognize the key characteristics of each. Few research papers ever spell out the scientific method used. Such things are meant to be intrinsically clear, but often they are not.

1.7 Ecology – where to in the future?

In the 100+ years that ecology has existed as a distinct discipline, we have made a lot of progress in understanding natural systems. Nevertheless, ecology has few general models that allow us to make correct predictions – generally considered to be a yardstick of scientific success. There are many reasons for this: ecology is still a young science, for example, and the inherent variability in most natural ecosystems, which change constantly in space and time, have posed great barriers to collecting data (a topic which we discuss further in Chapter 2).

Nevertheless, we must admit that ecology has a number of persistent weaknesses. Some areas of ecology are riddled with jargon, often seen as the hallmark of a poor science that uses the naming of things as a substitute for critically examining ideas (Peters 1991). Because of the explosion of literature, theories may be developed simultaneously in different parts of the literature (often using different jargon or names for the same model!) without any cross-reference. Relatively few ecologists work in a variety of habitats or with markedly different sorts of organisms, shifts that might encourage more cross-fertilization between different sections of the literature. Additionally, ecologists have been guilty of making vague, inaccurate, qualitative, subjective or inconsequential predictions – when they make any at all (Peters 1991).

All of the above problems, however, can be overcome, and we may hope that the next phase of ecology will see many of these problems largely corrected. This leaves us with the issue of whether there are general, predictive theories to be discovered in ecology. In the most pessimistic outlook, ecological systems could be so variable and intrinsically different from each other that nothing learned in one ecosystem will ever provide us with a model that is truly applicable elsewhere. If that is true, then ecology will never be anything more than a series of 'special cases'. We believe that such pessimism is not warranted, and that the future for ecological science is bright. Triumphs in science do not go to those who focus endlessly upon the differences between things, but to those who perceive similarities and can show us a new way of thinking about things. Recent understanding of some of the sources of variability in ecosystems is one such advance, a topic we discuss in the next chapter.

1.8 Further reading

Beveridge W.I.B. (1957) *The Art of Scientific Discovery*, 3rd edn. Heinemann, London.

Chalmers A.F. (1999) *What is This Thing Called Science?*, 3rd edn. University of Queensland Press, St Lucia, Queensland, Australia.

Gould S.J. (1981) *The Mismeasure of Man*. W.W. Norton, New York.

Hubbard R. (1990) *The Politics of Women's Biology*. Rutgers University Press, New Jersey.

Kirk R.E. (1995) *Experimental Design: Procedures for the Behavioral Sciences*, 3rd edn. Brooks/Cole Publishing, Pacific Grove, California.

McIntosh R.P. (1985) *The Background of Ecology: Concept and Theory*. Cambridge University Press, Cambridge.

Medawar P. (1982) *Pluto's Republic*. Oxford University Press, Oxford.

Peters R.H. (1991) *A Critique for Ecology*. Cambridge University Press, Cambridge.

Platt J.R. (1964) Strong inference. *Science* **146**, 347–353.

Underwood A.J. (1990) Experiments in ecology and management: their logics, functions and interpretations. *Australian Journal of Ecology* **15**, 365–389.

Chapter 2   *How does scale of measurement affect what we see?*

2.1   Introduction

In most lakes around the world small animals known as zooplankton live in the water column often near the surface. Many of these animals move slowly through the water grazing upon small plants, phytoplankton that also live in the surface waters, taking advantage of the often warmer temperatures and good sunlight. At Lake Waihola, a shallow coastal lake in Otago, New Zealand, a school group arrives to take some samples of the plankton. They have brought with them two plankton nets, one with an opening of $100\,cm^2$ and one with an opening of $2500\,cm^2$ (Fig. 2.1). They attach long ropes to each net and, standing on a small jetty, throw the nets out onto the lake in parallel, pulling them back through the surface waters. Plankton collected by each net ends up in a small bottle at its base. The students tip each bottle into a separate sample jar (along with a label recording the location, date, net size and names of the students). They then take their samples back to their laboratory and learn how to identify and count the different planktonic species. In the lab, the students are surprised to find that there are just as many individuals of a zooplankton species, the copepod *Boeckella triarticulata*, in samples collected by the small plankton net as there are in samples from the large net – even though the latter net sampled a much larger volume of water. Given that numbers of *Boeckella* are expressed as densities (that is, numbers of *Boeckella* in a sample are divided by the total volume of water sampled to calculate numbers of animals per cubic metre of water), the small-net sample suggests that *Boeckella* is far more abundant than do the large-net samples.

What has happened here? How can we get two very different estimates of density of the same species, in the same place, just by changing the

How does scale of measurement affect what we see?

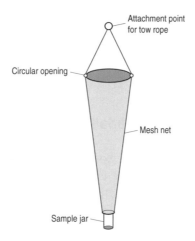

Fig. 2.1 A diagram of a plankton net (not to scale). Plankton nets are long cones that have circular openings at both the mouth and the apex, the latter having a sample jar attached, which can be removed after sampling. Nets come with different sizes of openings and mesh dimensions. The former characteristic affects the volume of water that is sampled and the latter determines the size of organisms retained by the net and collected into the sample jar. Plankton nets are pulled through the water for a set distance so that the total volume of water sampled (which equals the area of the mouth of the net multiplied by towing distance) can be calculated and kept consistent between replicate tows. Note that nets designed to sample benthic organisms work on similar principles and differ only in details of size and construction and use of area to bed to express densities rather than volume of water.

size of sampling equipment used? A possible explanation lies in the relationship between the size of the net and the distribution of zooplankton. If zooplankton occur in dense swarms of about the size of the opening of the small net, then it is possible that the larger net simultaneously sampled water that contained both very high and very low densities of zooplankton. In contrast, if the smaller net was by chance dragged through the middle of a zooplankton swarm, then it only sampled water containing a very high density of zooplankton.

Patchiness vs. the scale of sampling instruments is critical. In the preceding example, the results altered because the scale of measurement – in this case, the size of the sample nets – was changed. The observed difference is not real but is an artefact of slightly different scales of sampling. Virtually everything we measure in the environment has such **scale dependency**. Ecologists have known of this problem for a long time, but it is only recently that we have truly begun to recognize that not only must all measurements be properly scaled – that is, expressed in units of time and space – but that solving ecological problems means understanding how things change at different scales.

19

Understanding scaling is another basic requirement for the ecologist's intellectual toolbox. In this chapter, we introduce the basic problems of scaling. We begin by providing some basic definitions of scale and then go on to consider what is meant by scale dependency and how it can affect the conclusions we will draw.

2.2 A quick discussion of variables

Recall that in the last chapter (see Section 1.4.1) we described some of the questions that ecologists ask. These questions mean we measure things like numbers of individuals present (or perhaps percentage cover of the substratum in the case of colonial organisms – Box 2.1), or numbers of different species present, or numbers of organisms that perform a particular function (e.g. photosynthesis). We call these sorts of measurements **variables**, because their values vary at different times and places. Usually we need to measure these variables by collecting samples (see Box 1.5). Few organisms can be counted reliably with the naked eye, because many species are too small to be seen or are cryptic or secretive – hence we typically have to use sampling equipment to collect them. In freshwater environments, the most basic piece of equipment is a net, either for sampling benthic species or planktonic species. The exact form of the net we use will vary depending on the taxa we wish to sample, but a common theme is that the net opening samples either a known volume of water (for planktonic species) or a known area of the bed (for benthic organisms). Consequently, counts of organisms within each sample are expressed as densities – numbers per unit volume or unit area for planktonic and benthic organisms, respectively. This means we have the first way in which scale intrudes – we have imposed a scale of measurement through use of a piece of sampling equipment that collects organisms from a fixed volume or area of the environment. A second way in which scale of measurement is important is that we will have to decide the locations of boundaries within which we will take our

---

Box 2.1

Solitary and colonial organisms

In some animals and plants, the concept of separate, individual organisms does not really apply. Animals like sponges and corals and some plants grow as a **colony** of individuals that are intimately linked together so that individuals are subunits of that colony. This means that the fate of an 'individual'

depends on the fate of the colony in which it lives. For these organisms, it makes no sense to measure the population by the number of individuals. Instead, ecologists either count whole colonies (where these can be easily distinguished) or measure the percentage of the substrate covered by the organism, as many of these organisms are also sessile (attached permanently to one spot).

---

samples. For example, ecologists usually select **sites** in freshwater environments – particular lengths of river or locations on a lake (like a jetty)
from which samples will be collected (Fig. 2.2). The spatial sizes of sites
(for example, a 50-m long by 10-m wide length of river) also impose a
scale of measurement. Finally, we can look at the location of the site
within a region or even a continent. The geographical location of the
site might make a big difference.

We have now described three ways in which a sample is spatially
contingent, but it is important to realize that each sample is also temporally contingent in analogous ways. Each sample takes a particular length
of time to gather in the field – it may take seconds or minutes (in the case
of many net samples) or it may take a much longer time. We may choose
to collect all our samples within different **periods** of a day or a week or a
year – that is, we set the boundaries in time (like the boundaries of a site)
for sampling. Finally, we may sample at particular times of year (say
particular seasons) and of course we have to sample within particular
years that might be different from others – consider the sorts of differences that can be caused by the El Niño climatic effect, for example,
which rolls around about once every 5 years. So, for any set of samples
there are at least three sorts of scales we automatically impose by sampling – size of individual observations (in space and time), size of sampling sites and periods (in space and time, respectively), and location of

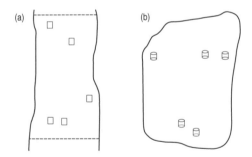

Fig. 2.2 An illustration of sampling scales (not to scale). (a) A section of river has been
selected as the sampling site (marked by the upper and lower dashed lines). Five samples of
organisms living within defined areas of the bed have been collected using a net (marked by
the square boxes) at points selected at random throughout the area demarcated by the
dashed lines. (b) A whole lake. Five plankton samples have been collected at random
points throughout the lake (marked by the cylinders). The two diagrams illustrate
different sampling scales. The river has been sampled within arbitrarily defined
boundaries of a site – consequently, the range of sampling is less than that of the whole
river, whereas the lake samples were collected from anywhere within the lake, which thus
represents the range of sampling. The river has been sampled using unit areas (because the
organisms being collected live on the surface of the bed) whereas the lake has been sampled
using unit volumes, because plankton live in the water column.

sites and sampling periods in larger geographical scales of space and longer scales of time, respectively. Let us define these scales, and then explore how and why changing the scales of measurement can change the patterns we see.

2.3   What is scale?

We follow the definitions of Schneider (1994), who found the misunderstanding of scale so rife in much of ecology that he wrote a whole book on the subject. First, the size of our individual samples is called the sampling **resolution** (or **grain**). Samples have a resolution in space (the actual area or volume of an individual sample) and a resolution in time (the time taken to complete an individual sample). For example, one benthic sample might sample $0.25\,m^2$ of streambed and be completed within 10 s. Second, the total area (or total volume) and duration of time over which samples are collected is called the **range** (or **extent**). Once again, sampling has a range in space – represented by the total area or volume falling within the boundaries of the site where we collect our samples – and a range in time – represented by the total time falling between the start and end of sampling. We can now define scale: **scale** is defined as the resolution within the range of a measured quantity, and we use this definition to identify the scale of a study in both space and time.

One other useful definition for measured quantities is that of **scope**, which is defined as the ratio of the range to the resolution. Hence sampling scope in space is the total area or volume in which samples were collected divided by the size of unit samples. Thus, if the size of our individual samples was $0.25\,m^2$ and we had collected our samples throughout an area of $10 \times 50\,m$ or $500\ m^2$, then the scope would be $500/0.25 = 2000$. Note that scopes have no units because the units cancel out. We can also calculate scope in time. If the duration of an individual sample is 10 s and our sampling is completed in 8 h, then the scope $= (8 \times 60 \times 60)/10 = 28800/10 = 2880$. Note that we have converted hours to seconds for the range, to keep the unit consistent with that of the resolution. As with any equation, you must make sure that different terms within it are expressed in the same units of measurement.

Why calculate the scope of a sampling programme? Scope is a ratio – its value tells us something about the relative sizes of the range and resolution. A very large number means that the total area examined (or the total duration of time) was very big compared to the size of the sampling instrument used to examine it. Because scale of measurement affects what we will see, the scope may alert us that the scale of a study was not appropriate, as we shall discuss further below. Sampling scopes are useful also in comparing different studies. For example, suppose we have a second survey in which the same sample unit size was used but samples were collected over 10×500 m or 5000 m^2 resulting in a spatial scope of 20 000. This survey has a scope 10 times larger than the one described above. It demonstrates clearly that the scales of the two studies are quite different. We can of course simply look at the ranges and resolutions without actually calculating the ratios, but the values of scopes can be quite useful in putting a number on the actual differences in scale between two studies.

We can also calculate the scope of particular phenomena that interest us. The **scope of a natural phenomenon** is simply the ratio of its upper (maximum outer scale) to its lower (least inner scale) limit. For example, suppose we are interested in the phenomenon of grazing of tiny, single-celled algae by an insect species in a stream. The lower limit here is the area occupied by a single individual of the algae (given that the grazer will be larger in size). Suppose the algae are typically about 200 μm long by 30 μm wide or about 6000 μm^2 in area – this is equivalent to 0.006 mm^2 (1 μm^2 = 0.000001 mm^2). The outer limit is set by the geographical range of the insect species. To keep things simple, let us consider individuals occurring in just one river system. Suppose we determine that the insect lives along some 50 km of channels with an average width of about 10 m, resulting in an approximate range of 50 000 m $\times$ 10 m = 500 000 m^2. We can now calculate the scope of grazing: it is 500 000 m^2/0.006 mm^2 or 500 000 m^2/0.000000006 m^2 (remember we have to correct the units), which is 8.33×10^{13} (or 833 with 11 zeroes after it). This is an enormous number! What it signals to us is that the array of scales over which grazing by the insect may affect the algae is huge. The significance of this number is particularly sobering when we compare it to the typical scope of an experiment designed to examine the effects of grazers on algae. Such experiments might have sample units of 100 cm^2 with samples collected over some 100 m of river channel, or perhaps 1000 m^2, to give a scope of 100 000 (or 10^6) or a fraction of the scales over which grazing occurs. Such mismatches are very common in ecology, because we rarely have the ability to collect and process samples over all the scales that might be relevant. Nevertheless, large differences in scope are a warning that our sampling surveys or experiments have only examined a small and possibly unrepresentative fraction of the populations we are interested in and the scales over which they occur.

You might be getting some inklings about why changing the scale of measurement can change the results, but now that we are armed with some basic definitions, we can return specifically to the question posed in the introduction – **how** and **why** does our scale of measurement affect what we will see?

2.4 How does the scale of measurement affect what we see?

2.4.1 Effects of sample resolution

What might happen if we change the resolution of our sample? Examine Fig. 2.3 where there are three proposed sample unit sizes (or resolutions) to be used over the illustrated 'landscape'. Sample unit areas like these are termed **quadrats**. Suppose our intrepid researcher is interested in finding out the density of organisms throughout this landscape to estimate population size. The first effect of differing resolution is that the estimate of organism density is likely to be quite different depending upon which resolution she uses. To understand this, imagine the researcher places the smallest quadrat at eight randomly selected locations throughout the landscape and counts the number of individuals in each sample. This small quadrat is likely to fall frequently within places where there are no or very few organisms and get only one or two places with a lot. This is because the organisms are not distributed in a uniform or even random way across the landscape, but are aggregated into small groups.

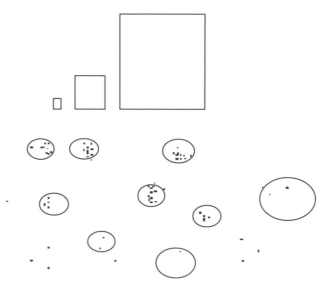

Fig. 2.3 Illustration of how changing the sampling resolution can change the patterns observed. The small black dots are individuals of a species to be sampled, the ovals are food patches with which (unbeknownst to the researcher) the species is spatially associated, although some individuals are between patches and some patches are unoccupied. The three squares are three hypothetical sampling unit sizes, each of which might potentially produce very different estimates of the mean densities of the species. The middle quadrat is also more likely to garner samples in which relatively high numbers of organisms and high food density are sampled together thus demonstrating the positive association. Note that only a portion of the landscape has been illustrated.

Consequently, the smallest sampling resolution will probably mean that the researcher ends up with a series of observations containing a large number of zeroes and one or two large numbers, perhaps on the order of 10 or 15 individuals. She can then calculate the **average density** by adding up the total number of individuals in the whole sample and dividing by the number of observations. If six of her observations were zeroes, one was 10 individuals and the eighth observation was 15 individuals, then the average would be $(10 + 15)/8 = 3.12$ individuals per unit area (let us call it area A). If we look at the landscape however, we might feel quite uneasy, because the statement that there are, on average, 3.12 individuals per unit A seems to convey little about the actual distribution and abundance of the target organism. Indeed this unease is quite reasonable because we are clearly sampling frequently in places where the organism does not occur much and then occasionally hitting an area where it is very dense. Simply taking an average across those two sorts of observations is a little like clarifying the use of dinner table condiments by explaining an average of the taste of pepper and salt – it really does not convey much meaning about either flavour! We can in fact express this unease formally as a lack of confidence in the sampling mean (Box 2.2).

This situation has arisen because of the size of the sampling unit *relative* to the size of aggregations of our target organism across the landscape. If we increase the sampling unit size by using either of the other two resolutions, the incidence of zeroes among the sample observations will decline because we will less frequently sample an area containing no individuals at all – especially with the largest quadrat. There will be less variation among individual observations within our sample, which will increase our confidence that the sample mean density is likely to be close

---

Box 2.2

The meaning of means

Most people are familiar with the formula for calculating an **average** (or **mean**, to give it its technical term), where we add up the values of our observations and divide by the total number of observations.

In sampling, we use sample observations to tell us about the rest of the population that we did not examine (see Box 1.5). As long as we have abided by the rules of sampling theory, a **sample mean** provides an unbiased estimate of the **population mean**, which is unknown to us. That is why sampling theory is so clever – but how well does the sample mean estimate the population mean? Sampling theory shows us that if there is a large range in values among the individual observations, and if we have only a small number of observations, then it is fairly likely that our sample mean is quite different from the population mean. We will not discuss how here, but we can calculate the amount of difference among sample observations and use this to tell us whether the sample mean is likely to estimate the population mean well or poorly.

You may not strictly understand this until you learn basic statistics, but there is a simple message you can learn now: *never, ever, ever* believe ('have confidence in') a sample mean someone shows you, unless they also provide an estimate of the variation among the observations used to calculate it!

to the true, and unknown, mean density over the whole landscape. This then is the first lesson about scale: our confidence in the sample mean can change with the size of the sampling unit we use, and this depends partly on the scales over which organisms are dispersed over the landscape.

You might be thinking at this point that the example seems a bit contrived, but keep a couple of things in mind. First, we cannot see, with the naked eye, the vast majority of organisms we sample, like we can in Fig. 2.3 where the clumped patterns are fairly obvious. Many species are too small, are cryptic or move too quickly for us to count them and see where they are located. Even large organisms that stay still, like large plants, pose a problem – we literally cannot see the forest for the trees. Second, many ecologists still select their sample resolution quite arbitrarily. Particular pieces of equipment are designed to sample fixed areas like $0.25\,m^2$ simply because that is what researchers have always traditionally used. Ecologists have known for a long time that resolution affects their confidence in sampling means, but it seems to have had very little impact on some sampling methods. As a consequence, it is common for ecologists to use a sampling resolution picked for reasons unrelated to the organism under study and to end up with a highly variable set of observations. In many cases, the latter is probably caused at least partly by the former.

There is a second way that sampling resolution affects our conclusions. Suppose that our researcher is interested in whether the distribution of this species is related to the amount of food in the environment. Of course, we can see in Fig. 2.3 that the target species is associated with the presence of food, but that the food is distributed in patches, the sizes of which are similar to the intermediate spatial resolution. If our researcher uses the biggest sampling resolution illustrated in Fig. 2.3, individual observations are likely to sample clumps of organisms and their associated food patch but also surrounding areas with lower densities and less food. However, none of these within-observation differences are registered, because we generate one value for numbers of individuals and one number for amount of food for each quadrat. A possible result is that relatively similar numbers of organisms and amounts of food are recorded in each observation. This would mean that our researcher might find no particular relation or only a very weak relation between abundance of organisms and amount of food. However, as we can see the association is quite obvious – but at a different spatial resolution. A sampling survey using the intermediate resolution had a better chance of detecting it. It is disconcerting, but common, in ecology for an association between two variables to be detected at one spatial resolution but vanish when the environment is sampled at another. Sampling resolution affects not only our confidence in sample means; it also affects our ability to detect interesting patterns in nature.

Before we leave sampling resolution, we should remember that each of the problems described above for spatial resolution have analogous problems for temporal resolution. That is, the time taken to collect an individual sample also has the potential to affect our confidence in sampling means or to detect patterns in nature. Suppose there are aggregations of individuals at particular times (as can occur when individuals converge to a particular spot for breeding, e.g. spawning salmon). If our sampling resolution in time is small relative to the duration of the aggregations then we have the same problem as before – we might get many zeroes accompanied by a few large numbers when we happened to sample at a time when many individuals were present. We might wish to take an average to calculate the abundance of the population over the duration of the sampling period, but might feel uneasy about the variation among individual observations. Likewise, if we increase the temporal resolution, we integrate across some of this temporal variability in numbers and potentially reduce the within-sample variability – but it may mean also that patterns we are seeking (such as the relation between numbers of individuals and a variable fluctuating through time) are lost to us.

How should ecologists determine the 'correct' sampling resolution? This is determined by the question we are asking, and whether we have some knowledge about the scales that are important. In the example above, if our researcher knew what spatial scales food varied over, she would have selected the intermediate resolution. When we have no information however, we may need to do quite detailed surveys initially (perhaps using an array of different resolutions) to examine how our organisms vary in space and time.

2.4.2 Effects of sample range

Range of sampling also has the potential to change our results. To see this, consider Fig. 2.4, where we have drawn the distribution of the same species as in Fig. 2.3 but over a greater overall range. The previous sampling area is located on the left, and with greater spatial perspective we can now see that the pattern of distribution and overall abundance suggested by that area is not consistent through space. If our sampling range was increased to include the area on the right, the average sample density would go down. If we did not record the fact that numbers were lower on one side we might just see relatively high variability among our observations and be unaware that we had sampled across some sort of important gradient. If a physical factor (like dissolved oxygen concentration) was responsible and we had measured that as well, then we might detect and understand the shift in abundances across our site. If the change in numbers is caused by something else – like the behaviour and distribution

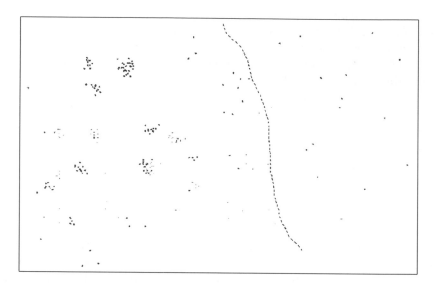

Fig. 2.4 Another landscape illustrating the distribution of the same species as in Fig. 2.3 but at a greater spatial range; the sampling range in Fig. 2.3 is toward the left. We can see that toward the right the distribution of the species in space has become more random and it is overall less dense. Such patterns could be caused by a physical gradient that makes persistence in the right-hand area less likely but also by biotic interactions. For example, the dashed line could mark the distributional limit of a predator that preys on our target species and is very successful at locating clumps of individuals.

of a predator that we did not know anything about – then we could end up with a gradient in numbers without understanding what causes it.

Once again, the example may seem contrived, but large-scale environmental gradients in physical conditions or in biological abundances are extraordinarily common in nature. As above, the problem is that we often cannot 'see' many of these gradients just by looking and hence put the boundaries of our sites in 'sensible' places. The analogous problem also exists with temporal range – the boundaries of sampling periods may be set in ways that are incongruous with temporal fluctuations in numbers. You would probably think then that ecologists are remarkably careful about sampling range, but sampling ranges (especially the boundaries of sampling 'sites') are often set as arbitrarily as sampling resolution.

2.5 A big problem: scaling up and down

The above discussion illustrates that virtually everything we measure in nature is likely to be scale dependent – that is, to alter in some way if we change the range or resolution. We define a **scale-dependent pattern** as

---

Box 2.3

Self-similar environments?

The physical structure of habitats (i.e. whether they are physically complex or simple) has big effects on organism density and diversity. Are there any simple rules for understanding how habitat complexity changes with scale that might allow us to understand shifts in species diversity at different scales?

Quite possibly – and you can test this for yourself. Take a map of a piece of coastline (that has no identifying features), show it to friends and ask them to guess what length of coastline they are looking at. Unless they happen to be familiar with that specific piece of coast, they will not be able to guess whether they are looking at 1000 m of coastline or 1000 km! This is because coastlines look the same at all spatial scales, a pattern pointed out in a fascinating paper by Mandelbrot (1967). Objects that have the same structure at different scales are called **self-similar**.

If environments are similar in physical structure at different scales, do plants and animals responding to that physical structure also scale accordingly? If so, would we be able to do experiments at small scales and confidently scale up to the large scales where we cannot collect data because it is logistically too difficult? This is an active research area!

---

one where some measure of the pattern changes if either the resolution or range of measurement is changed. This then brings us to one of the biggest obstacles in ecology: if patterns (and processes that determine them, such as interactions between species) are scale dependent, we cannot extrapolate our data to scales we have not measured. To see why this is such an obstacle, recall the example about the scope of an algae–grazer interaction discussed previously compared to the scope of a typical experiment looking at the interaction. The mismatch in scales is huge, hence we cannot take the results of that small-scale experiment and confidently scale them up to the whole river system. We usually have no idea, for example, how the interaction might change at other 'sites' along the river channel, so we could not be confident that the results would apply elsewhere. We cannot simply multiply up results (such as average densities per unit area of sampled substratum) to calculate densities over the entire river channel (or other scales we have not measured) and expect that these estimates will bear much resemblance to the actual numbers of organisms.

The challenge for ecologists now is to discover scaling rules – formulae that would provide the understanding that would allow us to scale up or scale down our results confidently. Scaling rules could exist, for example, if there are common ways in which species shift in numbers across particular scales – perhaps because the environment shifts in predictable ways across scales (there are some hints that this may be the case – see Box 2.3). Discovery of such rules could revolutionize the way we do ecology. Until then, the best advice is to take extreme care with our choices of scale and interpretation of data!

2.6 Further reading

Levin, S.A. (1992) The problem of pattern and scale in ecology: The Robert H. MacArthur Award Lecture. *Ecology* **73**, 1943–1967.

Mandelbrot B. (1967) How long is the coast of Britain? Statistical self-similarity and fractional dimensions. *Science* **156**, 636–638.

Mandelbrot B. (1983) *The Fractal Geometry of Nature*. W.H. Freeman & Co., San Francisco.

Schneider D.C. (1994) *Quantitative Ecology: Spatial and Temporal Scaling*. Academic Press, San Diego.

Wiens J.A. (1989) Spatial scaling in ecology. *Functional Ecology* **3**, 385–397.

Those readers interested in self-similar landscapes might try reading *Chaos: Making a New Science* by James Gleick (Penguin, London, 1987), which includes some neat pictures generated from mathematics.

Chapter 3   *Wonderful water: linkages from the atom to the biosphere*

3.1   A day in the life of a water molecule

... You have no idea what sort of a day I have had today! There I was, swirling with my friends in the tropical ocean off Fiji, admiring all those multi-coloured fish when I felt the currents pass me up towards the ocean's surface. Not a problem. But then I realized that the hydrogen bonds between me and my friends seemed to be growing weaker. Things were moving fast and chaotic, and before I knew it, I felt I was flying in space.

You're right – evaporation! But not in any straight line upwards – oh no. It was like a crazy ride in all directions, one moment close to a bunch of other molecules and next, far away. It was like living two lives as water and as vapour. As I got higher, it got cooler and I seemed to spend more time in the water phase. I thought it pretty odd because water ought to be denser than air. So I was not surprised that as our droplet of water accumulated more molecules (and some pretty agitated static electricity – those electrons sure know how to party!), we began to fall a little. There were lots of false starts – we'd begin to fall and then get sucked up even higher to form a larger droplet. But then, when we were all feeling pretty cold and slow, it began to rain properly and we started the downward trip.

As I fell, I realized that I was not falling back into the ocean I had left but was going to splash down onto the ground somewhere. We all started saying our goodbyes because we knew that the droplet was likely to be broken up again. That's one thing about being a water molecule – you are always forming hydrogen bonds with new molecules and you never seem to share the same electrons with the same molecule.

With a thud we landed, along with lots of other droplets. I saw some landing on the trees, sliding down the trunks as stemflow, and being funnelled down to the roots. I actually landed on a flowing little stream of droplets that had fallen and amalgamated earlier, and we started to flow downhill. By now, the rain had stopped and all around me I could see other molecules being dragged down in among the sand grains to become part of the groundwater. I had been there before, quite a few years ago so I was in no hurry to join them.

We soon stopped flowing downhill, ending up in what seemed to be a puddle with some tadpoles in it. There were lots of other particles among the water molecules – mostly bits of clay, covered in phosphorus atoms and other elements. At least there weren't too many heavy metals around – one pool I fell in once seemed to be full of bad elements!

So here I am. And as I remember the multi-coloured fish in Fiji, I have a nasty suspicion that this tadpole swimming towards me is going to . . . its mouth is opening . . . arghhhh, I feel another adventure coming on . . .

3.2   Water, water, everywhere . . .

Perhaps the story above is a little dramatic but even so, we are not dealing with any ordinary substance when we look at this deceptively simple-looking molecule – water. In the rest of this chapter, we shall explore the fine-scale molecular linkages among the atoms in water that govern the global linkages among water stores in the hydrological cycle. We shall see how water's chemical properties affect life in, on and near water. Perhaps most importantly, we shall see how these same properties render water vulnerable to some types of pollution and other human impacts.

Water is a familiar substance to us all. There is no known life form on earth that does not depend on water for at least part of its lifecycle. Water in various solutions acts as the medium for many chemical reactions, a transport agent for circulating material around plants and animals, and the hospitable environment for sperm and eggs to fuse in sexually reproducing organisms. A continuous ocean of water covers 70% of the planet, the atmosphere includes a protective blanket of water vapour constantly changing in volume and distribution while below the land's surface, a large fraction comprises the groundwater and soil water in terrestrial ecosystems.

Over half of our body weight is water. A typical 70-kg person has some 42 L of water inside their body. In their cells are 23 L while another 19 L swills around outside, including 3 L in blood plasma, 9 L in the lymph and interstitial spaces, 6 L in connective tissue (e.g. bones and tendons) and skin, and 1 L in joint spaces and around the brain and spinal cord. This water is far from static. Isotope studies have shown that water molecules in the capillaries (tiny blood vessels) of the human arm exchange across the capillary walls some 300 times per minute. Within 3 h, isotopically labelled water may be completely distributed throughout the human body (Robinson 1978).

3.3   Why is water such a remarkable molecule?

Up until near the end of the 18th century, water was considered an element, and it took the genius of the French chemist Lavoisier to interpret the results of Cavendish's experiments and show that water is actually composed of oxygen and hydrogen. Working out the exact formula took some time after that, but now any school-child can tell you that water consists of two hydrogen atoms and an oxygen atom. The molecule is extraordinarily stable, needing a massive amount of energy to break it into hydrogen and oxygen. But for all its familiarity and apparent chemical simplicity, the molecule is truly remarkable and has unexpected properties that control virtually all physical, chemical and biological processes occurring in lakes and rivers worldwide.

Chemists call water a 'hydride of oxygen'. Unlike other hydrides of elements in the same chemical group of the Periodic Table as oxygen (e.g. H_2Se, H_2S), it is a liquid between 0 and 100°C at standard atmospheric pressure instead of a gas. It also freezes at a much higher temperature than would be expected. In solids, molecules are held in a reasonably constant relationship to each other – individual molecules vibrate but about a fixed position. Conversely, gas molecules move rapidly and randomly with no structural relationship to each other. Liquids are harder to describe. Are they highly condensed gases where the molecules still move randomly but are closer to each other or are they disturbed solids where molecular arrangements are loose but still important?

Fluid water is considered to be a 'liquid crystal'. Most of water's anomalous properties result from the fact that the two hydrogen atoms are held at an angle of 104°27' (Fig. 3.1). Powerful covalent bonds link the atoms but although the molecule is electrically neutral overall ($2H^+ + O^{-2}$), the oxygen nucleus has a slightly higher affinity for electrons than those of the hydrogen atoms. This 'polar' nature means that the slightly positive H atom of one molecule is attracted to the slightly negative O atom of an adjacent molecule to link them through a 'hydrogen bond' (Fig. 3.1). Thus, at a micro-scale, these two types of chemical linkages (covalent and hydrogen bonds) hold the molecules in a tetrahedral structure that, in solid form, we know as ice.

This crystal is quite open, with the oxygen atom surrounded by four neighbouring hydrogen atoms. As a result of its 'openness', ice floats which is crucial for many organisms living in large lakes and rivers in cold areas (see Chapter 4). By forming a floating insulating layer, the water below the ice layer does not freeze solid so that many fish and

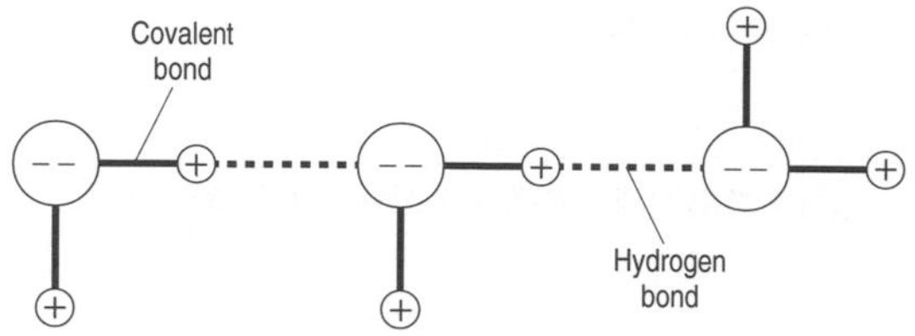

Fig. 3.1 The bonds within (covalent) and between (hydrogen bonds) water molecules are responsible for many of water's unusual chemical and physical properties. The negatively charged oxygen atom is shown by the larger open circle.

other organisms can survive over the cold season (as long as the oxygen levels remain high enough). However, some invertebrates are able to tolerate complete freezing. For example, in some high-altitude shallow Canadian ponds that freeze totally, damselfly nymphs survive the winter frozen in the ice (Daborn 1971).

In many small streams in cold regions, ice crystals falling from frozen, overhanging vegetation can form nuclei for frazil ice that accumulates in spiky masses on any solid object in the water. Over time, this can become anchor ice that may dam and divert water out of the streambed. Apparently, fish and insects become disoriented by the changes in flow and loss of recognizable objects in the stream, wandering into the diverted channels. When the ice dam melts and water returns to the normal channel, many aquatic animals are stranded and die. In high mountain streams such as Sagehen Creek in Sierra Nevada, California, ice dams are considered the major source of mortality in young trout and overwintering insects (Needham & Jones 1959).

3.4 Ecologically relevant properties of water

At temperatures and pressures on the earth's surface, water can exist in all three phases: solid, liquid and gas. Each of these three phases is significant to life on the planet but we shall restrict ourselves to considering the liquid aspects here. Virtually all the ecologically relevant properties of water described below arise due to the linkages (hydrogen bonds) among water molecules. But these properties also result from the particular environmental conditions on earth – what water exists on Venus is all vapour whereas only ice has been found on Mars. In a way, the unique properties of water combine with the unique conditions on earth to provide the perfect planetary panoply for such diversity of aquatic ecosystems and their processes (Philip 1978).

(a)

(b)

Fig. 3.2 (a) A temporary pond on the Ovens River floodplain in south-east Australia. Considerable daily fluctuation in air temperature may occur, yet water temperatures vary over only a range of 3°C. (b) Lake Te Anau, on the South Island of New Zealand, is over 70 km long and over 300 m deep and formed following the retreat of ice-age glaciers. Lake water temperature shows little annual fluctuation despite considerable variation in annual air temperature due to the buffering effect of the massive volume of water present in the lake.

Water plays a crucial role in buffering variations in temperature. Because of its relatively high **specific heat**, it takes about five times as much heat to raise a given mass of water by 1°C as it does to warm the same mass of dry soil through the same temperature range. Thus, the water temperature of even a shallow lake may vary little over a daily cycle, ranging between 18 and 20°C, whilst air temperatures vary from 12 to 28°C. Temperatures in larger lakes vary even less, with little variation in temperatures being observed, even over annual cycles (Fig. 3.2).

Water also has a high latent heat of evaporation (i.e. the amount of heat required to evaporate a liquid) because the crystal structure of water must be completely broken down. However, water evaporates at standard atmospheric pressure at temperatures below 100°C as long as the water content of the air in contact with the water surface is below saturation. This evaporation process enables plants to transport fluid up their stems when they transpire and lose water through their stomata. The latent heat of evaporation provides a 'cooling effect' on damp surfaces (e.g. sweaty skin) as well as the upper layers of lakes, sometimes leading to thermal stratification (see Chapter 4).

The hydrogen bond linkages make water moderately viscous. A pressure difference (usually a gradient down a slope) is needed to start water moving and keep it flowing, overcoming friction. Viscosity generates the turbulence of fast flows over rough surfaces in streams, a crucial source of oxygenation of water (see Chapter 5) and an agent in creating a variety of microhabitats for aquatic life. Second only to mercury, water has the highest surface tension of any liquid on the earth's surface, again due to the hydrogen bonds. Such high surface tension allows capillary forces to wet soil above the water table without filling all the interstitial cracks. This is essential for plant roots that require not only water but also gases. The air–water interface is a special habitat created by the high surface tension of water. Above, we can find floating plants, microbial films or agile insects that scurry on the surface film (Williams & Feltmate 1992). Below the meniscus, there may be organisms such as mosquito larvae that hang down, piercing the layer with tubes to breathe air. Other forms such as flatworms or snails crawl slowly upside-down as if on a solid surface. We shall see later that all these organisms comprise a distinct assemblage termed the neuston (see Chapter 4).

3.5 How does water circulate at a global scale?

At the other end of the spectrum from the micro-scale of molecular linkages lies macro-scale of the **hydrological cycle** and the linkages among the global **water compartments**, that is rivers, lakes, oceans, and so on (Fig. 3.3). The hydrological cycle is the continuous circulation of water between the earth and the atmosphere and through the various compartments. The cycle is powered by gravity and solar energy. Gravitational influences are obvious in many of the vertical pathways such as the fall of rain, stemflow and the downhill movement of running waters. Direct solar effects include the sun's energy for the linkages of evaporation and transpiration. Solar energy also influences the circulation of weather cells, generates winds and affects climatic variability. The relative contributions of the linkages in the hydrological cycle change over time and across the earth's surface, and together with human influences, these variations in the hydrological cycle govern the global distribution and water regime of our standing and running waters.

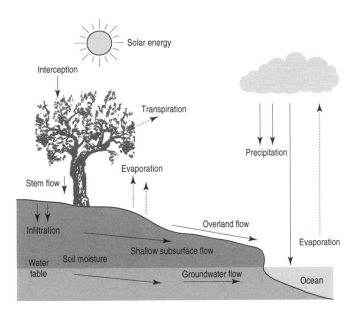

Fig. 3.3 A simplified diagram of the hydrological cycle, illustrating the linkages and pathways of water among many global compartments. Some aspects have not been included (e.g. ice-cap melting, capillary movement in soils).

Across the globe, evaporation from the oceans ($505\,000\,\text{km}^3$) exceeds the $458\,000\,\text{km}^3$ of precipitation that falls on them. Some $119\,000\,\text{km}^3$ precipitation falls on land (less than one-third of the earth's surface) and $72\,000\,\text{km}^3$ returns by evaporation to the atmosphere. The difference ($47\,000\,\text{km}^3$) eventually runs off as surface or groundwater. The existence of lakes and rivers on the land masses of the world arise because of this spatial 'imbalance' in the water budget of the hydrological cycle.

Estimates of the **volumes** of water in the various compartments of the global hydrological cycle vary slightly, but the relative magnitudes are the same. Most of the planet's water is present in the oceans (94%, Table 3.1), with another 4% in the groundwater. At any one time, the water contained within the lakes and rivers that are the topic of this book comprises less than 0.02% of the global water budget (Table 3.1). However, whilst the amount of water in lakes and rivers at any point in time is small relative to larger compartments such as the oceans, the amount of water passing through lakes and rivers over time is substantial. The water in most lakes and certainly all rivers is constantly draining out of the system, and hence must be replaced by inflowing water if the system is to be maintained. We express the average amount of time a water molecule will remain within a system as the **residence time**. Streams and rivers may have residence times ranging from only days to months,

Table 3.1 The principal compartments, volumes, percentages and mean residence times of global water stores. (After Boulton & Brock 1999.)

Compartment	Volume ($1000\,km^3$)	Per cent	Mean residence time (years)
Oceans	1370 000	94.202	3000
Groundwater	60 000	4.126	5000
(Actively exchanging groundwater)	(4000)	(0.275)	(300)
Glaciers/ice caps	24 000	1.650	8600
Lakes/inland seas	230	0.0158	10
Soil water	82	0.0056	1
Atmospheric vapour	14	0.00096	0.027
Rivers	1.2	0.00008	0.032

whereas large lakes and inland seas typically have residence times of years to decades (Table 3.1). In contrast, oceans and groundwater systems have water residence times of 3000–5000 years and hence complete water turnover rates are very slow.

The volumes and residence times of water within the compartments of the hydrological cycle are as important as the linkages among them. If we put this in practical terms, we would expect a relationship between residence time of water in a particular compartment and its vulnerability to anthropogenic pollution. Rivers are likely to be more susceptible to pollution than glaciers or the ocean. The high rates of water flow through rivers (compared with glaciers) increases the likelihood of polluted water flowing into a river at some point in time. Once a pollutant does find its way into a river, the relatively small volume of a river (relative to an ocean) means that only limited dilution of the pollutant will occur. However, there is no scope for complacency with respect to the risks associated with pollution of systems with long residence times. When a pollutant enters such a system, then it is likely to stay in that system for a long time. For example, groundwater pollution is widespread and chronic, and in many cases will persist for many hundreds of years, even after the source of the pollution is removed. Pollution also spreads more slowly in systems with long residence times, hence substantial quantities of the pollutant may enter the system before the problem is actually detected (Winter *et al.* 1998).

3.6 Human impacts on the hydrological cycle

Later in this book, we shall review some of the local effects of humans and their activities on the organisms and ecological processes occurring in standing and running waters in various parts of the world. However, there are global trends in humans' effects on the hydrological cycle that are important to consider. One such effect is **climate change**, largely resulting from changes in greenhouse gas concentrations (e.g. carbon dioxide, methane, nitrous oxide) that have been rising since the industrial age. Globally since 1750, atmospheric carbon dioxide has risen by 30%, methane by 145% and nitrous oxide by 15% largely due to fossil fuel use, agriculture and changes in land use (Houghton *et al*. 1996). The rates of such increases are forecast to accelerate.

The overall trend is predicted to be global warming, with associated changes in evaporation and precipitation patterns. While changes in temperature will affect hydrology, most experts consider the main impacts will be on annual and seasonal amounts of rainfall and its variability and intensity. There are many models being used to predict changes at global, continental and regional levels but generally scientists agree that climatic change is occurring and that extreme events (drought and flooding) are likely to become more frequent. In regions where small changes in rainfall lead to proportionally larger changes to runoff (e.g. most semi-arid areas), effects on the regional hydrological cycle are expected to be marked (Grimm *et al*. 1997). However, a lack of long-term data and the confounding effects of numerous other human activities make it impossible to single out the effect of global warming.

Despite the prominence given to global warming issues and the obvious effects on rainfall and runoff in the hydrological cycle, the effects of modification of catchment vegetation may be quantitatively more significant (Smith 1998). Most human land-use activities entail clearing trees and deep-rooted vegetation from the catchment and replacing them with pastures or urban areas. This primarily alters water tables and groundwater movements, water uptake and transpiration rates, and the amounts and variability of runoff.

Direct impacts on some of the pathways of the hydrological cycle include dams on river systems that alter the residence time of runoff. Conversely, urbanization, land clearance and river channelization have all increased runoff rates in other areas. Inter-basin transfers and the extraction of water from surface waters and groundwater also distort the linkages and volumes of compartments regionally and collectively. The cumulative effects of numerous small dams on the upper reaches of many river systems can alter the local hydrological cycle substantially. All of these activities affect the natural water regimes of standing and running waters. As well as changing the physical features of the hydrological cycle, there are also severe ecological impacts of these alterations to natural water regimes (see Chapter 10).

At a global scale, do we have enough fresh water to meet our needs into the future? Estimates vary widely but most prognoses are gloomy. For example, Postel *et al*. (1996) calculate that humanity presently uses 26% of total terrestrial evapotranspiration and 54% of the world's accessible runoff. Evapotranspiration is the combination of evaporation and transpiration, and represents the water supply for all non-irrigated vegetation. We probably could not increase our use of evapotranspiration because most land suitable for rain-fed agriculture is already being used. In the next 30 years, new dams could increase accessible runoff by about 10% whereas the population is projected to increase by more than 45%. The global water future is not bright. It is estimated that one-third of the world's population is living under moderate or severe water stress, especially in the Middle East and North Africa and approximately 1.3 billion people lack access to adequate supplies of safe water.

Furthermore, the issue is not only **water quantity** in the hydrological cycle but **water quality**. While we may be able to make some adjustments to our impact on the hydrological cycle through increased efficiency of water use and changes in agriculture, we should also simultaneously minimize our pollution of presently available freshwater (Postel *et al*. 1996). The same linkages that maintain the dynamic flows of the hydrological cycle can also convey pollutants and contaminants (see Chapter 11). Our management of global water resources must acknowledge the central role of these linkages, and understand how these linkages affect ecological processes.

3.7 Further reading

Daborn G.R. (1971) Survival and mortality of coenagrionid nymphs (Odonata: Zygoptera) from the ice of an aestival pond. *Canadian Journal of Zoology* **49**, 569–571.

Grimm N.B., Chacon A., Dahm C.N. *et al*. (1997) Sensitivity of aquatic ecosystems to climatic and anthropogenic changes: the Basin and Range, American Southwest and Mexico. *Hydrological Processes* **11**, 1023–1041.

Houghton J.T., Meira Filho L.G., Callender B.A., Harris N., Kattenberg A. & Maskell K. (eds) (1996) *Climate Change 1995: The Science of Climate Change*. Cambridge University Press, Cambridge.

Needham P.R. & Jones A.C. (1959) Flow, temperature, solar radiation and ice in relation to activities of fishes in Sagehen Creek, California. *Ecology* **40**, 465–474.

Philip J.R. (1978) Water on Earth. In: McIntyre A.K. (ed.) *Water: Planets, Plants and People*. Australian Academy of Science, Canberra, pp. 35–59.

Postel S.L., Daily G.C. & Ehrlich P.R. (1996) Human appropriation of renewable fresh water. *Science* **217**, 785–788.

Robinson J.R. (1978) Water in the animal body. In: McIntyre A.K. (ed.) *Water: Planets, Plants and People*. Australian Academy of Science, Canberra, pp. 60–70.

Smith D.I. (1998) *Water in Australia. Resources and Management*. Oxford University Press, Melbourne.

Williams D.D. & Feltmate B.W. (1992) *Aquatic Insects*. CAB International, Wallingford.

Winter T.C., Harvey J.W., Franke O.L. & Alley W.M. (1998) *Ground Water and Surface Water – A Single Resource*. United States Geological Survey Circular 1139, Denver, Colorado.

Chapter 4   *What is in lakes?*

4.1   A year in the life of a temperate lake in Wisconsin

It is mid-winter in Wisconsin. Warmly dressed ice-skaters gyrate across the surface of a frozen lake, unaware of the extraordinary processes occurring in the cold, clear water beneath them. Fortunately for the skaters, ice floats, whereas water is at its most dense near 4°C. If we cut a hole in the ice and measure the water temperature at various depths, we find that the lake is **stratified**, with a cool, dense layer of water of 4°C at the bottom and a cooler but less dense layer floating just below the ice (Fig. 4.1a; see Chapter 3). The ice prevents the wind from mixing the cold layer just below and insulates the underlying water. If ice were heavier than water, it would sink to the bottom, and the entire lake would freeze solid from the bottom up.

The ice melts in spring, and as the surface of the lake warms, the layers mix (Fig. 4.1b). The upper layer warms, becomes denser (near 4°C) and sinks. With only small differences in temperature and hence water density between the surface and bottom water of the lake, it only takes the slightest breeze over the surface of the lake to turn over the water column and completely mix the waters of the lake. Complete spring mixing of the layers occurs over a short period of time, but the result can be dramatic. The combination of warming and light from above and resuspension of vital nutrients from the lower layer produce ideal conditions for the proliferation of certain types of tiny suspended plants (phytoplankton). A spring 'bloom' of these organisms results, and for a while, there may be several million of these tiny organisms in a single litre of water. The lake turns a murky green.

As spring progresses, the weather warms further. Surprisingly quickly (sometimes in a few days), stratification occurs again, this time with a warm upper layer (e.g. 18–20°C) overlying cooler, denser water (12–14°C, Fig. 4.1c). Access to the critical nutrients from the lower layer is

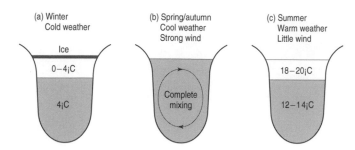

Fig. 4.1 Patterns of stratification and mixing and associated weather conditions in a small dimictic Wisconsin lake. Dimictic lakes stratify twice a year, under ice in winter and when surface layers warm in summer.

cut off by the stratification, limiting the daily growth of phytoplankton. It does not take long for grazers and parasites to mount an attack on the phytoplankton bloom and soon the spring bloom has finished. The lake once again becomes relatively clear.

It is mid-summer. Standing in a punt in the middle of the lake, two swimmers with snorkels and flippers dive over the side and swim as deep as they can. At about 6 m, they pass quite suddenly from the warm surface water into the much cooler water of the now stratified lake. Even though quite strong winds may blow in summer, the stratification is stable, maintained by the density differences in water of different temperatures. Within this lake it is almost as if two distinct ecosystems have formed, separated by a thin layer where water temperature changes sharply, that is until the next turnover in autumn when surface temperatures fall.

An obvious question arises – do differences in the properties of the water layers at different depths in the lake influence the communities of plants and animals that live in them? If such questions form in our mind as we dive down through the layers of the lake, then we have taken the first step of beginning a scientific investigation into understanding how lake communities function.

4.2 Seeking patterns in the diversity of standing waters: describing and categorizing lakes

Science is all about prediction. We have just looked at a year in the life of one lake. Based on our observations and our knowledge of the properties of water (see Chapter 3), we can begin to speculate as to the factors that might influence the dynamics of living organisms in that lake. If we understand one lake, then we might be able to predict how other lakes will behave. However, is this what happens in all lakes? Every year? Probably not, given that standing or lentic waters encompass a huge variety of habitats, ranging from small, temporary ponds that fill for a few short weeks every few years in the deserts of the Australian outback to vast permanent lakes such as the Great Lakes of North America.

A possible solution to dealing with the diversity of different lakes is to seek some feature or a combination of features to group similar lakes. Possible features that we could use include depth, surface area, water clarity or stratification. Having just read the description of the Wisconsin lake, the description of a lake as being 'stratified' will now tell you something about a range of conditions that are likely to be encountered in a lake that stratifies. It also gives a clue as to how we might start to compare different lakes. Do all lakes stratify? If not, why is it that some lakes stratify and others do not?

Hutchinson (1957) proposed a system of lake classification based on stratification that is still used widely by limnologists today. The first distinction is between **amictic** lakes that never mix (e.g. some Antarctic lakes), **holomictic** lakes that mix totally, and **meromictic** lakes that mix incompletely (see below). Holomictic lakes can then be further divided into those that mix twice a year like the ice-covered Wisconsin lake above and those that mix only once a year, usually only stratifying in summer. A third group represents those that mix more than twice a year, stratifying briefly during warm, still weather before being mixed by strong winds, a cycle that may occur numerous times over the course of a year in some climatic regions. Meromictic lakes are typically very deep lakes or those receiving inflows of high concentrations of solutes (Fig. 4.2a). Water rich in dissolved material is denser and sinks to the bottom of the basin. Where this happens, it can form a permanent deep layer that never mixes, unless disrupted by some spectacular event (for example, the 'exploding' Lake Nyos in Cameroon; Kling *et al.* 1987, Kling 1987). Several distinct water layers can occasionally form in meromictic lakes. We can get a permanent bottom stratum due to **chemical** density differences below more than one temporary upper strata formed by **thermal** density differences (Fig. 4.2b).

This functional classification leads us to ask how lake basin shape and climate might influence stratification. If strong relationships can be found, maybe we can predict the ecological processes that will occur in certain lakes. Try this for a lake near you. If you live near a deep lake sheltered from the wind by a valley or tall vegetation in a temperate climate, chances are the lake stratifies once a year, or may even be permanently stratified (less common). If it gets covered in ice, it probably stratifies twice. If, on the other hand, the lake is quite shallow, clear and often whipped by wind-generated waves, it may not stratify or, if so, only briefly during warm lulls in weather.

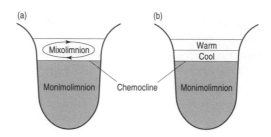

Fig. 4.2 Patterns of stratification in meromictic lakes. The deeper monimolimnion (a) never mixes due to chemical stratification. Thermal stratification of the surface mixolimnion (b) may occur in summer due to surface warming.

4.3 What are the physical and chemical consequences of stratification?

Stratification affects nearly every process in the lake. When light enters water, it is absorbed rapidly and exponentially with depth. We would expect heat to follow the same trend but in fact it does not, largely because wind mixes the upper layers of water, distributing the absorbed heat. Thus, stratification causes 'layering' of water temperature (see Fig. 4.1). As many chemical and biological processes are controlled by temperature, the rates of these differ between the two strata. Further, the absence of any mixing means that the chemical composition (solutes, dissolved gases, pH, etc.) of the water in the two different layers can be very different. In permanently stratified lakes, the differences are even more marked and in the bottom stratum, termed the **monimolimnion** (see Fig. 4.2), environmental conditions compared with surface layers are truly worlds apart.

Mixing of dissolved gases is severely restricted between the strata. As a result, the upper layer is usually well-oxygenated while the lower layer becomes hypoxic (low in oxygen). In the monimolimnion, the highly soluble 'rotten-egg gas' (hydrogen sulfide, H_2S) may build up to extraordinary concentrations. While surface water usually has no odour, water from the depths often stinks because of H_2S. The black mud found in such deep hypoxic places (giving the Black Sea its name) results from chemical transformations of sulfur into ferrous sulfide, and is a sure sign of little or no oxygen.

In the well-lit upper layer (the **euphotic** zone – Fig. 4.3), photosynthetic plants and microbes use solar energy and carbon dioxide to generate their body tissues (see below) but consume dissolved nutrients as they grow. This shortage of nutrients often limits the growth or yield of these organisms. On the other hand, nutrient concentrations may be quite

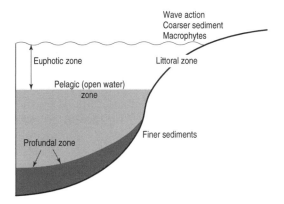

Fig. 4.3 The four major habitats in standing waters. The euphotic zone indicates the depth to which light sufficient to sustain photosynthesis penetrates.

high in the lower layer. When mixing occurs, the nutrients present in this deeper water become temporarily available to the surface producers resulting in algal blooms.

The acidity of the water usually differs between the layers. We express acidity as pH (the concentration of hydrogen ions); the lower the pH value, the more acidic the water. Water in the lower layers is usually more acidic than at the surface although acidity can change in response to high rates of photosynthesis. Acidity affects the solubility of most dissolved metals, ions and nutrients, and hence their accessibility to aquatic organisms.

4.4 Where to live? Habitats in standing waters

Looking out across a lake, we can also begin to identify areas of the lake that are quite different from each other (Fig. 4.3). If our lake is shallow, we may be able to see the bottom of the entire lake, and conditions may not vary much from the edge to the centre. In contrast, if the lake is deep, the bottom will probably only be visible around the lake margins, an area known as the **littoral zone** (Fig. 4.3). The substrate may vary according to the size and exposure of the shore to wind and waves. If you are standing on the shore of a large lake where large waves may be generated on windy days, then it is likely that the substrate will consist of coarse sand and cobbles. If the lake is small and sheltered, then it is probable that the substrate will consist of fine mud. Aquatic plants may also be abundant here. If we refer back to our descriptions of lake stratification, water in the littoral zone is usually warm and well oxygenated, at least in summer and during the day. However, low oxygen conditions can occur at night when photosynthesis ceases or where large amounts of organic material accumulate and begin to decay. Such accumulations may occur where wind-driven material piles up or where inflowing streams deposit material from the surrounding catchment.

If the lake is shallow (say less than 1 m deep), well lit and clear, we find that warm and well-oxygenated conditions may prevail right across the lake. However, if the lake is deep, then conditions on the bed of the lake will rapidly change as we go deeper. Light will quickly diminish with depth, ultimately declining to a point where photosynthesis and hence the growth of plants and algae is impossible. The depth at which photosynthesis ceases to occur varies greatly, although lakes in which any significant photosynthesis occurs at depths greater than 30 m are exceedingly rare. This deep and dark region of the lake is known as the **profundal zone** (Fig. 4.3). At such depths, wave action no longer disturbs the water column. In the absence of any major water movement, fine loose inorganic and organic sediments will cover much of the bed of the lake. Lack of any mixing of either the water column or the sediments may mean that the lake sediments become **hypoxic** (low in oxygen), and any organisms that live here will require adaptations for survival in low oxygen environments. Hypoxic sediments will almost certainly occur if the lake is either holomictic or meromictic.

Away from the bed of the lake, the open waters of the lake are known as the **pelagic zone** (Fig. 4.3). If the lake is large and deep, then the pelagic zone may be a vast, three-dimensional space. However, based on our prior knowledge of patterns of lake mixing, we now know that this space is unlikely to be featureless. Even in a well-mixed, non-stratified lake, considerable changes in the prevailing conditions can be expected with increasing depth. The upper part of the pelagic zone will be well lit, hence photosynthesis can occur. However, rates of photosynthesis are highly dependent on the availability of nutrients and other factors such as turbidity (see Section 4.5.1) and rates of zooplankton grazing (see Chapter 12). As depth increases, rates of photosynthesis decline and respiration becomes the dominant process. Organisms inhabiting such deep water environments will primarily depend on the breakdown of organic matter for energy. As discussed previously, if the lake is stratified then very sudden changes in a variety of physical and chemical conditions will occur with increasing depth (see Section 4.3). Such changes will have important implications for the organisms that can potentially live in the pelagic zone and the depths at which we might expect to find them.

Knowing the patterns of physical and chemical variation that occur within a lake allows us to speculate on the nature of the **habitats** (places where organisms live) that are available. If we also have some idea of the types of organisms that typically live in lakes and ponds, then we can also begin to hypothesize as to what organisms are likely to live in any part of the lake or pond that we might care to sample. The interactions between the various physical and chemical features of lakes described previously form the stage upon which organisms can act out their lives.

4.5   What lives in lakes?

The organisms living in any habitat represent a subset of organisms selected from a potentially larger set living in a wider area. In our previous example, the range of species inhabiting the Wisconsin lake will be drawn from the pool of species that inhabits lakes in that part of North America. Which species are actually found in our Wisconsin lake will depend on a range of factors. Two key determinants are (i) the physical and chemical conditions within our lake and the tolerances of different species to those conditions in the local species pool; and (ii) the abilities of different species to disperse into our lake. To be in the position where we can develop meaningful hypotheses regarding what animals we are likely to find in the different habitats of a lake and what they do there, we need to have some idea of the typical organisms that occur in lakes.

There are likely to be hundreds or even thousands of different types of organisms living in any large lake. Again, a first step in attempting to understand their diversity is classification. In most lakes, the two most conspicuous groups of organisms are the plants and animals. Plants are generally **photosynthetic**; that is, they build their bodies by harvesting energy from the sun. Animals are **heterotrophic** and build their bodies using energy obtained by eating other plants and animals. An additional, less conspicuous element of the biota comprises **microbes** – a hugely diverse assemblage of microscopic organisms that may be either photosynthetic or heterotrophic.

4.5.1   Plants

Macrophytes or large multicellular plants in lakes may be classified into several groups based on whether they are rooted (attached to a substrate) or floating, and whether all parts remain submerged or some parts emerge out of the water (emergent) (Fig. 4.4). Based on our knowledge of lake habitats, it is not hard to hypothesize how the growth form of a particular plant might determine where it can grow. Plants are photosynthetic and therefore need access to light in order to grow. However, the intensity of light fades rapidly with increasing depth so plants must grow near the surface. Water clarity will influence how far

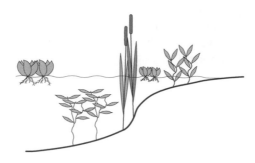

Fig. 4.4 Some patterns of macrophyte growth in lentic habitats. From left, floating, submerged, tall emergent, littoral floating and shallow water emergent.

light penetrates into the water, and hence determine how deep plants can grow. In very turbid lakes, there may be insufficient light to support plant growth at depths greater than 1 m. In very clear lakes, plants may be present on the lake bed at depths greater than 20 m. Rooted emergent plants generally grow in water less than 10 m deep, and hence are most commonly found in the shallow, sloping littoral zones of lakes. In very shallow lakes, they may cover extensive areas leaving only limited areas of open water. Floating plants can grow anywhere on a lake as they are not restricted by a lack of light. However, they are vulnerable to disturbance from wind and wave action which can leave them washed up high and dry on the shore so they usually are restricted to either small sheltered lakes or in the protected zones between the stems of larger emergent water plants.

4.5.2 Microbial organisms

The microbial organisms present in most lakes include unicellular and colonial algae, rotifers, protozoa, bacteria and blue–green algae. Although individuals may be invisible to the naked eye, their combined biomass may be huge. Collectively, they can have profound impacts on the physical and chemical conditions in a lake, playing key roles in nutrient cycling, primary production and the decomposition of organic material. The accumulations of decaying organic detritus that may be found in sheltered areas such as the bottom of lakes or amongst macrophyte stems often support rich microbial communities and form a key food resource for many invertebrates and fish (see Section 4.6). Various microbial organisms grow as a **biofilm** (living layer) over most hard substrates within a lake. In well-lit areas, these biofilms are frequently dominated by various photosynthetic algae and are a food for grazing invertebrates such as snails. In low nutrient lakes with clear water (**oligotrophic** lakes), primary production by algae growing in biofilms in the littoral zone may be the primary source of energy for larger heterotrophic organisms such as snails and mayfly nymphs (see Section 4.5.3).

Organisms that float freely through the water column are collectively termed **plankton**. Their movements may be entirely passive, drifting wherever the currents take them. Alternatively, they may have some capacity to control the depth at which they occur, although their ability to control horizontal movement can be limited. Photosynthetic algae that float freely within the pelagic zone form the **phytoplankton** (*phyto* = plant). Phytoplankton are a diverse array of organisms ranging from microscopic, unicellular blue–green algae to colonial *Volvox* that form pale green balls that can be clearly seen with the naked eye. Phytoplankton is the primary food for many species of filter-feeding **zooplankton** (*zoo* = animal) (see Section 4.5.3). Dense blooms of phytoplankton may occur in lakes where nutrients are abundant, turning the lake turbid and green (a **eutrophic** lake – also see Chapter 12). Little light can penetrate to the bottom under such conditions, thus preventing the growth of benthic macrophytes and photosynthetic algae in biofilms. As a result, phytoplankton may form the basis of the food web in eutrophic lakes (see Section 4.6). A wide variety of other heterotrophic microbial organisms, including unicellular bacteria, ciliate and flagellate protozoa and multicellular rotifers, may also be abundant in the pelagic environment. Such organisms can form an important link between photosynthetic organisms and larger invertebrate grazers (see Section 4.6).

4.5.3 Invertebrates

Animals can be conveniently divided into invertebrates and vertebrates: those animals without and with well-defined backbones, respectively. Various types of crustaceans, molluscs, oligochaete worms and the adults, larvae or nymphs of insects usually dominate the invertebrate fauna of lakes, although a wide variety of other invertebrates can be encountered. The tolerance of different invertebrates to extreme physical and chemical conditions varies widely, although many invertebrates can tolerate far more severe conditions than vertebrates such as fish. Most vertebrates are excluded from habitats that are low in dissolved oxygen although dense populations of various invertebrates specialized for survival under hypoxic conditions may flourish. Such specializations may include numerous gills, or perhaps abundant haemoglobin that turns these organisms' bodies bright red. Such 'bloodworms' can often be found in abundance in the hypoxic mud at the bottom of lakes and include the larvae of various midge flies (chironomids) and oligochaete worms. The ability of invertebrates to tolerate extreme physical and chemical conditions combined with their relatively small size and their often considerable capacity for aerial dispersal means that even the most isolated or environmentally extreme lake will support some invertebrate species.

The shapes of different invertebrates often reveal their habitat preferences. **Benthic** (bottom dwelling) invertebrates tend to be either flattened or worm-like in shape if they burrow through sediments. Oligochaete worms burrow through the sediments present on the bottom of lakes, their shape allowing them to slip past substrate particles and through mats of macrophyte roots and organic debris. Alternatively, well-developed legs such as those found in crayfish and shrimps or the large muscular foot possessed by molluscs, allow the animals to hold and move over various substrates. In well-lit areas where macrophyte beds flourish, molluscs such as gastropod snails often dominate, grazing the prolific growths of algae that often cover any hard surface. Other bivalve molluscs, such as the zebra mussel (see Chapter 13) filter microscopic animals from the water. Aggressive, predatory insects, such as damselfly nymphs, may be common on hard substrates such as weeds and woody debris. Abundances of the more conspicuous invertebrates (such as damselfly nymphs) and small fish are often particularly high in macrophyte beds relative to other areas, presumably due to the presence of food and protection afforded by the dense macrophyte cover from larger predatory fish.

Out in the open waters of the pelagic zone, animals have to be able to swim or float to maintain position. Long limbs or antennae propel a variety of crustaceans and insects through the water. Most zooplankton are small crustaceans less than 5 mm long, such as many species of calanoid copepods and cladocera (water fleas). Many zooplankton are filter feeders, consuming a wide variety of tiny unicellular organisms that inhabit the pelagic zone. The larger cladocera, including many species of *Daphnia*, are particularly efficient filter feeders and can have a significant influence on water clarity and quality (see Chapter 12). In contrast, other species of zooplankton are predatory and eat smaller zooplankton, a relatively common feeding strategy amongst various copepods.

The small size of many zooplankton species means that their swimming abilities are relatively weak and they may find themselves carried considerable distances by wind-generated currents and waves. However, although their capacity to swim is often limited, their distribution within the water column is unlikely to be random. Different species of zooplankton are frequently found in discrete water layers within the pelagic zone, or alternatively discrete littoral habitats at certain times of the day. Many species may migrate over considerable vertical and horizontal distances over a 24-h cycle. Whilst some species rise vertically into the surface waters at night, others may sink down into deeper water layers. Species that spend the day in fringing littoral macrophyte beds may migrate horizontally out into the pelagic zone at night. Many zooplankton species form well-defined aggregations (often called 'swarms') of varying size in the pelagic zone. Various explanations have been suggested for these vertical and horizontal movements and the formation of swarms, including avoidance of predators, selection of optimum water temperatures, tracking of food resources and mate searching (Folt & Burns 1999).

Larger invertebrates may be able to swim well and are thus better able to orientate themselves in the water column. This group of invertebrates is termed the **nekton**. The nekton includes crustaceans that are active pelagic predators such as mysid shrimps, and insects such as backswimmers (notonectids) and diving beetles (dytiscid beetles). In many cases they actively seek and consume smaller zooplankton and benthic invertebrates. Phantom midge larvae (*Chaoborus*) regulate their vertical position in the water column using gas-filled hydrostatic organs, and may migrate over vertical distances of greater than 10 m. A further group of animals termed the **neuston** live in association with the surface of the water. Most neustonic animals are insects, but there are some spiders. Long-legged water striders can run across the water's surface, pouncing on other small invertebrates that have the misfortune to fall in and are unable to escape the water's surface. Other predators, such as whirligig or gyrinid beetles, have paddle-shaped legs and propel themselves across the water's surface like tiny boats. Gyrinid beetles have two pairs of eyes, allowing them to simultaneously scan both the air above and the water below for potential danger or food. Many of the insect members of the nekton and neuston are also very capable flyers, and can actively move between waterbodies (Williams & Feltmate 1992). For this reason, predatory insects such as backswimmers, diving beetles and water striders are often widely distributed, readily colonizing waterbodies that refill following floods or heavy rain (Lake *et al.* 1989).

4.5.4 Vertebrates

A variety of vertebrate animals live in and around lakes, and include fish, amphibians, reptiles, birds and mammals. Whilst fish may be found in virtually any well-oxygenated habitat within a lake, air-breathing reptiles, birds and mammals are generally restricted to the shallow littoral areas and surface layers of the pelagic zone. The distribution of amphibians is broader given that larval amphibians (e.g. tadpoles) have gills and breathe underwater. However, they are generally most common in shallow, littoral habitats where dense macrophyte beds provide them with abundant food and protection from predators such as fish. Tadpoles are often very rare or absent from lakes in which fish are present.

Fish are generally the most conspicuous, completely aquatic animals that live in lakes, although their distribution among lakes can vary depending on connections to other aquatic systems (Matthews 1998). Many isolated lakes contain no fish, with important consequences for the structure of these lake communities (see Chapter 12). Large, isolated fish-free lakes most frequently occur in high mountains or in the high-latitude areas of North America and Eurasia where depressions scoured by glacial ice during the last ice-age have subsequently filled with water. Fish are also frequently absent from small, often temporary, ponds due to isolation and the extreme physical and chemical conditions that may regularly occur. For example, fish are very rarely found in isolated ponds that either periodically dry or alternatively ice over and become deoxygenated in winter.

The bird fauna of lakes is particularly rich, with different species making use of all potential habitats in waters less than 10 m deep. Various species of bird browse vegetation (ducks and swans), filter zooplankton (flamingoes) or catch fish and macroinvertebrates (cormorants, grebes and sea eagles). Some birds stand and feed in shallow littoral waters, whereas others such as cormorants and grebes may dive several metres down to obtain food. Many birds such as ibis may feed elsewhere but breed or roost in and around lentic habitats. Because birds are highly mobile, they can arrive at newly inundated lakes within days of filling. In many floodplain lakes in arid, inland Australia, long dry periods are interspersed with short 'booms' of productivity after floods. During the 1990 flood when the normally dry central Australian Lake Eyre filled, an estimated 1 million waterbirds flocked to the newly-filled wetlands. Huge colonies of Australian pelicans, cormorants, black swans, terns and silver gulls were established in what was previously hot and dry lakebed sediments (Kingsford *et al.* 1999). No one knows how these birds 'sense' the water and migrate into the desert for long distances.

4.6 Lake food webs

Several different energy sources may form the base of lake food webs (Fig. 4.5). In clear, low nutrient, oligotrophic lakes, little primary production occurs in the water column and benthic production may form the primary energy base. The lake littoral zone may extend to a depth of over 20 m, and either macrophytes or benthic algae may capture the energy from sunlight. However, relatively few animals consume macrophytes directly. Hence even where macrophytes are abundant, benthic algae may still form the basis of the lake food web. Algae growing on hard substrates and the surface of macrophytes can be grazed by a variety of benthic invertebrates, particularly gastropod snails and decapod shrimps.

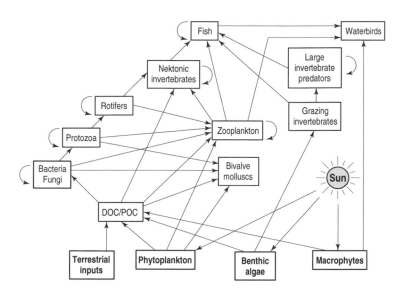

Fig. 4.5 A lake food web illustrating key pathways of energy flow in lakes. The direction of the arrow indicates the flow of energy along the link. Arrows that loop within a group indicate organisms within this group may prey upon each other. DOC/POC refers to dissolved organic carbon and particulate organic carbon, respectively. Note, although not indicated for simplicity, all consumer groups return material back to the DOC/POC compartment as either faeces or when they die and decompose.

In turbid eutrophic lakes, light penetration may be limited hence little benthic primary production occurs. In contrast, the production of phytoplankton may be significant. In such lakes, zooplankton may attain very high densities and filter phytoplankton from the water column. They can form a crucial link between microscopic primary producers and the larger predators, including zooplanktivorous invertebrates and fish (Fig. 4.5).

Bacteria also play a key role in transferring organic carbon fixed through primary production to the rest of the food web, a feature of lake food webs that has become known as the 'microbial loop'. A significant proportion of the organic carbon fixed by primary producers may be subsequently released as **dissolved organic carbon** (DOC) or **particulate organic carbon** (POC). Various bacteria and fungi utilize DOC/POC as an energy source, and in turn these bacteria and fungi can form a significant food resource for larger invertebrates. Bacteria and fungi also play crucial role in the decomposition of organic material. Whilst relatively few organisms consume macrophytes whilst they are alive, they can be an important food resource once they die. Dead macrophytes along with a wide variety of other sources of organic carbon derived from animal faeces and corpses, terrestrial plant litter and the remains of phytoplankton and zooplankton enter the pool of detritus in the lake. Detritus is a key resource for many organisms (Fig. 4.5). A wide variety of microbial organisms can colonize organic detritus, and complex microbial food webs may exist entirely within the detritus. Microbial primary producers are preyed upon by various bacteria and predatory protozoa (particularly ciliates). In turn, these microscopic predators are eaten by larger predators such as rotifers, nematodes and microcrustaceans such as harpacticoid copepods (Fig. 4.5).

Particularly complex food webs can occur amongst the diverse range of nektonic and benthic invertebrates that occupy lake littoral and benthic habitats (Fig. 4.5). Grazing invertebrates include gastropod snails. Detritivores include midge larvae and oligochaete worms. Feeding upon these generally non-predatory invertebrates is a wide variety of predatory invertebrates including the nymphs of damselflies and dragonflies, the adults and larvae of dytiscid water beetles, and many species of juvenile and adult fish. Large decapod crustaceans such as crayfish and shrimps may also be common in such habitats. Decapods are frequently omnivorous, feeding on detritus and algae much of the time, but will opportunistically switch to invertebrate prey when available.

4.7 What to study in lakes?

A knowledge of the common organisms that live in lakes, their typical habitats and patterns of feeding, along with the key physical and chemical features of a lake, allows us to start seeking answers to the questions that may have occurred to you through this chapter or perhaps when you last stood beside a pond or lake. Are all lakes the same, or can we group them into a number of similar kinds of lakes? Do changes to the physical and chemical conditions influence the distribution of the organisms that live in lakes? How does the structure of a lake community change over a year? Do the activities of predators alter the distribution of their prey? Sampling different habitats or lakes using a consistent methodology allows us to begin to tackle such questions. A bewildering variety of specialist sampling tools and methods exists and can be used to compare different lakes and habitats within lakes. Useful descriptions of sampling methods can be found in various texts (Merritt *et al.* 1996, Rabeni 1996). However, perhaps the most versatile item of equipment to start studying the ecology of a lake is a simple, long-handled dip net. It is the sampling technique that most lake ecologists first use, and continue to use at various times for their entire careers. Start with a simple question. For example, do the communities of invertebrates in a macrophyte bed differ from the invertebrate communities found in areas of open water? Compare the communities using a consistent, replicated sampling method (see Chapters 1 and 2), and you are doing ecological science.

4.8 Further reading

Dodds W.K. (2002) *Freshwater Ecology: Concepts and Environmental Applications.* Academic Press, San Diego.

Folt C.L. & Burns C.W. (1999) Biological drivers of zooplankton patchiness. *Trends in Ecology and Evolution* **14**, 300–305.

Hutchinson G.E. (1957) *A Treatise on Limnology*, Vol. 1. *Geography, Physics and Chemistry*. John Wiley and Sons, New York.

Kingsford R.T., Curtin A.L & Porter J. (1999) Water flows on Cooper Creek in arid Australia determine 'boom' and 'bust' periods for waterbirds. *Biological Conservation* **88**, 231–248.

Kling G.W. (1987) Seasonal mixing and catastrophic degassing in tropical lakes, Cameroon, West Africa. *Science* **237**, 1022–1024.

Kling G.W., Clark M.A., Compton H.R. *et al.* (1987) The 1986 Lake Nyos gas disaster in Cameroon, West Africa. *Science* **236**, 169–175.

Lake P.S., Bayly I.A.E. & Morton D.W. (1989) The phenology of a temporary pond in western Victoria, Australia, with special reference to invertebrate succession. *Archiv für Hydrobiologie* **115**, 171–202.

Matthews W.J. (1998) *Patterns in Freshwater Fish Ecology*. Chapman & Hall, New York.

Merritt R.W. & Cummins K.W. (1996) *An Introduction to the Aquatic Insects of North America*. Kendall/Hunt Publishing, Dubuque, Iowa.

Rabeni C.F. (1996) Invertebrates. In: Murphy B.R. & Willis D.W. (eds) *Fisheries Techniques*. American Fisheries Society, Bethesda, 335–352.

Williams D.D. & Feltmate B.W. (1992) *Aquatic Insects*. CAB International, Wallingford.

Chapter 5 *What is in rivers and streams?*

5.1 Contrasts among running waters

In the sandy streambed of the Kruis River, southern Africa, a shrinking pool full of dying aquatic insects attracts a hungry lizard. It is May but the austral winter rains have not yet commenced in earnest. However, an intense thunderstorm upstream has initiated a spate that cascades downstream, even as the lizard gulps down a mayfly nymph trapped on the surface scum. Around a corner, the water comes rapidly – not as a wall of swirling mud, but nonetheless shifting the unstable sands of the streambed and resuspending leaves and sticks to carry them downstream. In another kilometre, most of the energy is gone and the water seeps into the sand. A new braided channel exists briefly before drying out. The inhabitants of the shrinking pool, having been dispersed by the small spate, lie motionless or stagger on the dry bed to be eaten by terrestrial predators such as ants and lizards. A new pool formed by the spate lies in the lee of a tree trunk that has been half-buried in the channel. Already an adult mosquito hovers over it, ready to lay her eggs.

On the same day but on the other side of the world, a siren sounds repeatedly, warning of an imminent flow release down Miribel Canal, a bypassed channel of the Rhône River near Lyon. This canal, artificially straightened and stabilized, now conveys water from the hydroelectricity station, and its flow alternates rapidly from one stable discharge level to another. Gone are the enriching floods that used to sweep across the channel's floodplains, rejuvenating wetlands and distributing sediments and organic matter. The rapid rise and fall of water level in this regulated section has taken its toll on the aquatic inhabitants unable to avoid stranding. Backwaters and pools where floating water plants once thrived have gone, and the present channel complexity is nothing like its former self.

In both these contrasting systems, flow orchestrates the physical, chemical and biological processes that occur within the channel. In one river, a wide spectrum of flows moving along a winding channel generates a complex mosaic of habitats for aquatic and terrestrial organisms. In the other, human intervention has altered the flow regime and removed much of the channel complexity so that many natural processes either no longer occur or are impaired. Construction of dams and weirs have also altered migration pathways of fish and other animals while the constant rises and falls in water level create harsh conditions that only a relatively small number of species can survive.

5.2   Does the typical river exist?

The opening section contrasts two very different rivers, and illustrates the diversity of just two types of running waters. We face a problem similar to that faced with describing and classifying lakes (see Chapter 4). How can we measure, describe and attempt to classify and understand systems that are so variable across every level of spatial and temporal scale that we might wish to consider? Perhaps our first step might be to develop a descriptive model that incorporates many of the features seen in a moderately sized river system. From that descriptive model, we can go on to contrast our 'typical' river with other types of river systems.

Most people living in areas of moderate to high rainfall will have a similar mental image of a river system, no matter which continent or island they live on. Small, fast-flowing, rocky streams tumble down hillsides. As headwater streams merge, the size of the stream increases. Large slow-flowing sections form as the gradient lessens, and fine silt and organic matter accumulates on the bed of the stream. Further downstream, the size of the stream may have increased to such an extent that we now call it a river. Close to the sea, complex patterns of mixing of freshwater and seawater in an estuary may occur as inflowing tides push seawater upstream, often for quite considerable distances. A major change in the biota of the river may occur as only those organisms adapted to cope with sudden changes in salinity can survive in estuarine areas.

Whether the river system described above fits your preconceptions of what a typical river system looks like will probably depend on where you live. If you are reading this book in an arid region, then nearby rivers may only flow intermittently. Much of the time they probably consist of a series of still, isolated pools. In areas where high mountains have been carved by ice-age glaciers, fast-flowing rocky rivers may drop straight off abrupt cliffs and into marine fiords without passing through a slow-flowing lowland or estuarine section. Clearly, there is no such thing as the typical river system, and it seems reasonable to expect that the biota of any particular river system will vary with respect to the local patterns of flow, geology, geomorphology and myriad other factors that might structure a river community. However, despite this variation, we can begin to consider how the habitats and living organisms might vary with longitudinal changes in flow and channel structure.

5.3   Describing key features of river systems

Rivers are connected to all parts of their catchment by the movement of water. They are elongated ecosystems, carrying water, dissolved and particulate nutrients, organic matter, plants, microbes and animals down-hill. Sometimes they overflow, flooding surrounding land and exchanging material with the surrounding floodplain. Floods are critical events governing ecological processes both within the river and in adjacent ecosystems, highlighting the interconnected nature of ecosystems as discussed previously (see Chapter 3).

Rivers exhibit different patterns of complexity at various levels of spatial and temporal scale. Therefore, we must describe their many features at different scales. At the catchment level, we might use a map or aerial photograph to determine drainage density (length of channel per unit area of the catchment). We would also see the drainage patterns, and be able to tell something about the catchment geology from them. As branches of the headwater streams join, we would expect the channel to get larger and average discharge to increase. In everyday language, we consider 'streams' as small running waters with lower discharge than 'rivers' but like many common terms, one person's idea of a stream may not match that of another person.

One rapid method for estimating stream size is Strahler's stream ordering system where two 'first order' streams join to give a 'second order', two second orders to give a third, and so on (Fig. 5.1a). Another is Shreve's method that sums the orders of the upstream tributaries (Fig. 5.1b). Both these methods have their limitations (map scale, for a start!) but they can provide a rough idea of channel size and have been used to indicate distance along a river system when predicting longitudinal changes. However, be prepared for surprises. Rivers may occasionally decrease in size moving downstream, particularly if the river is dependent on the supply of water from headwater streams, and further downstream water is pumped from the river for town water supplies and irrigation.

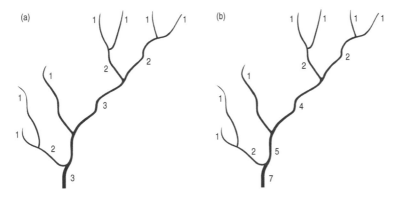

Fig. 5.1 Two ways of expressing stream size as 'stream order'. The Strahler method (a) is rapid but may not be useful for lowland rivers where many small tributaries join without changing stream order, whereas the Shreve method (b) overcomes this problem but is time-consuming with large rivers.

At a finer scale of metres to kilometres, we might focus on more specific channel features. Many of these are related to channel shape and provide some clues about the habitats available for organisms in the river. In upland streams, most channels are narrow, V-shaped and controlled largely by bedrock. The streambed comprises large rocks, cobbles and a little gravel. Downstream, where the gradient is more gradual, the river is likely to meander in a broad valley. Here the channel is controlled by alluvial (river-borne) sediment, the river is wider and the bed is more likely to be mainly sand, silt or clay.

5.4 How can flow organize processes in running waters?

In Chapter 4, we examined how ecologists use stratification as a key feature to help understand how lakes work. In lotic systems, flow is a key feature that we can use to classify rivers. Flowing water is what makes a river a river. Even in rivers where flow ceases for some of the time, the occurrence of flow, no matter how rare, will determine many of the system's key features.

Whether flow is continuous or intermittent is a good point at which to start developing a classification scheme. We can separate continuously flowing waters (**permanent**) from those that cease to flow either regularly (**intermittent**) or occasionally and unpredictably (**episodic**). Related to flow is discharge, the volume of water passing a point in a given time. Discharge (expressed as Q in cubic metres per second) is estimated as the average current velocity (m/s) multiplied by the cross-section of the stream (m^2).

The Tools of Freshwater Ecological Science

To describe flow patterns in running waters more accurately, we use **hydrographs**. These diagrams illustrate changes over time in water height (stage) or discharge at a point along the river, and reveal much about the river's behaviour. For example, look at Fig. 5.2. The first thing to notice is the time scale – we can see that these hydrographs reveal the flow regime of the rivers over multiple years so we are getting an idea of seasonal and annual flow patterns. Next, we look at the vertical scale of discharge and see that the annual discharge of the Cooper is much lower than the Mekong. Nevertheless, annual averages would obviously be misleading because the variation in annual discharge in the Cooper is immense compared to the highly regular annual flow regime of the Mekong. In the Cooper, some years there is no flow at all (at least at the gauging station!) and so we would expect the aquatic inhabitants to have behavioural or physiological adaptations to drying. On the other hand, the Mekong is permanent, and in response to a predictable seasonal climate, has a regular flow regime. The hydrograph of the Mekong River also resembles that from a highly regulated river

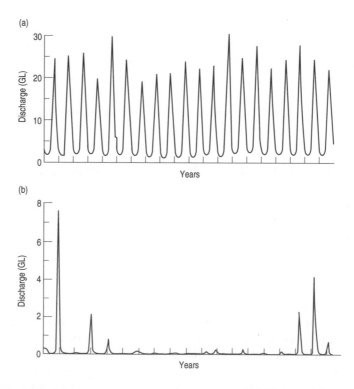

Fig. 5.2 Twenty-year hydrographs of two contrasting rivers: (a) the Mekong River (SE Asia), a predictable, tropical river; and (b) Cooper Creek (Central Australia), an irregular, arid-zone river.

60

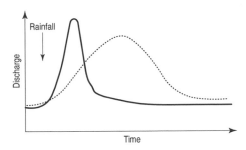

Fig. 5.3 A hydrograph depicting hypothetical patterns of discharge from urbanized (solid line) and vegetated (broken line) river catchments following a single rainfall event.

where water is released seasonally for irrigation. If such a seasonal pattern of regulation were imposed on the natural regime of the highly unpredictable Cooper Creek, the repercussions on the natural ecological processes in that system would most likely be disastrous.

We can also look at hydrographs measured at finer scales of hours or days, and this can tell us about runoff characteristics in response to heavy rainfall (Fig. 5.3). For example, a steep rising limb of discharge would indicate a sudden response, most likely due to rapid runoff from the catchment. This is common in urban streams where impervious surfaces like roads funnel water swiftly into the stream channel. In arid areas, lack of vegetation and the hard-baked ground can have the same effect, creating flash floods. The slope of the receding limb can indicate how much water continues to enter from upstream or drain from the catchment after rainfall has ceased. A shallow falling limb might indicate a densely vegetated catchment that releases water slowly as subsurface flow. It is also important to know where the gauging station is along the river – curves will usually be steeper in the upper reaches than lower sections of permanent rivers. The shape of the hydrograph influences many chemical and biological features of rivers including the dilution of river water or the stranding of organisms by receding floodwaters.

5.5 Flow, sediments and transport

Flowing water transports sediment. In general, the faster the current, the larger the particle that can be carried, although this depends on its shape, packing in the streambed, density and the type of flow. Flow in a stream channel may be **laminar** where parallel layers of water shear over each other smoothly or, more commonly, **turbulent** and mixing in a complex fashion. Turbulent flow buffets large sediment particles, sometimes rolling them along the bed. Smaller particles are carried along in the water column as suspended load. The extremely fine sediments that comprise clay may take an unexpectedly high current to move them as they are 'glued together' by strong electrical forces. However, once these fine particles of silt and clay are suspended in the water column, they may only need very low velocities to keep them moving and the resultant murky water is termed **turbid**. Many lowland rivers are naturally turbid, although poor management of streamside vegetation or excessive land clearance and siltation can also contribute to high turbidity. In turn, this reduces light entering the stream and may prevent some aquatic plants from photosynthesizing. Suspended sediment may also interfere with respiratory surfaces like fish gills. Fine sediment also acts as a carrier for various chemicals. Much of the phosphorus in running waters travels attached to suspended particles.

5.6 Chemical features of running waters

In many respects, the same processes that affect water chemistry in standing waters also apply in streams and rivers, e.g. diffusion of gases into the water is crucial for the supply of oxygen to aquatic organisms. However, many of these processes are also influenced by flow. Turbulent flow in upland streams often enhances the solution of dissolved gases such as oxygen, and concentrations may even exceed 100%. Moving water is also capable of eroding away rock material (weathering) and this can change the concentration of chemicals in the water. One example is the role of water in eroding limestone (calcium carbonate) where, because of the slight acidity caused by carbon dioxide dissolved in the water (producing dilute carbonic acid), the limestone is eaten away by the water creating intricate channels and cuttings in the rock.

Another example of how flow affects chemical processes in running waters is **nutrient spiralling**. In lakes, nutrients such as phosphorus may be cycled repeatedly between the water layers, plankton and sediment organisms as the lake alternately stratifies and mixes (Fig. 5.4a). Eventually, the nutrient may leave the lake, perhaps as some plant matter eaten by a duck before it flies away. If we superimpose the cycling nature of these nutrients onto the continuous flow in running waters, we get a downstream spiral (Fig. 5.4b). The nutrients are transported downstream, alternately being taken up by living organisms and then released as waste products or through the processes of death and decay. The tighter the links in the spiral, the more 'retentive' is the stream (Gomi *et al.* 2002). For example, in a natural stream, rocks and logs trap and retain organic material, slowing its progress downstream. This can provide various organisms with an opportunity to take up nutrients, further delaying the loss of nutrients downstream. However, an urban drainage gutter (a)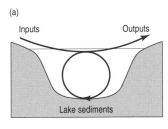

(b)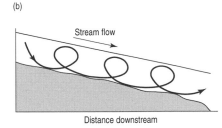

Fig. 5.4 Patterns of nutrient cycling and transport in (a) lakes and (b) rivers. In lakes, nutrients may be recycled repeatedly before ultimately being exported. In rivers, nutrients are transported downstream, although retention of nutrients within the stream may vary greatly between systems. Tighter spirals indicate that the stream retains nutrients more efficiently.

would have very loose spirals due to the poor retention of organic material and would be unlikely to support much life.

5.7 Where to live? Habitats in running waters

A feature of the distribution of biotic communities along river systems is longitudinal zonation. Microbial, plant and animal communities tend to exhibit progressive change as we move from headwaters downstream. As discussed above, differences between headwater streams and lowland rivers are not restricted only to size, but can span across a whole range of physical and chemical features. The longitudinal changes in community structure along a river system reflect differences in the capacity of various species to tolerate changing physical and chemical conditions, along with differences in food resources, predators and competitors along the system (see Section 5.9).

At a fine level of spatial and temporal scale, the interaction between flowing water and channel structure is a key factor determining where different organisms can live (Fig. 5.5). Turbulent patterns of flow down steep, upland channels create habitats that are spatially and temporally highly variable at relatively fine scales. Considerable variation in water velocity, substrate and potential food sources may occur over the scale of millimetres to metres. Habitats in such streams are often unstable, with the force of the water creating the potential for substrate movement. Variation in water velocity results in considerable variation in the composition of benthic substrates, often over small areas. Fine sediments and organic detritus tend to accumulate in areas of slow flow, whereas coarser cobbles and boulders dominate in faster-flowing areas. Other habitats, including macrophytes and woody debris, may occur

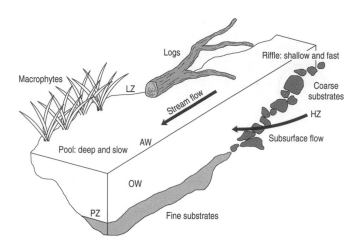

Fig. 5.5 Macro- and micro-habitats in running waters. Macro-habitats indicated are open water (OW), the littoral zone (LZ), the profundal zone (PZ) in large rivers, the hyporheic zone (HZ) where subsurface water exchanges with the surface stream, and the air–water interface (AW). A variety of micro-habitats exists across pools and riffles, and amongst different substrates. Arrows indicate direction of water flow.

along the stream channel. The ability to move between habitat patches in such an unstable, spatially heterogeneous environment can be advantageous. A diverse community of animals may also be found deep within the bed of the stream, a habitat known as the hyporheic zone. Stream water infiltrates into the **hyporheic zone** carrying with it particles of organic material, or exfiltrates from the hyporheic zone back into the stream. Photosynthesis is impossible in this zone, and the animals that live here will be either detritivores or predators.

In very large rivers, the sheer size of the channel combined with a reduced gradient can mean that relatively little variation in current velocity occurs across much of the riverbed. The lack of spatial and temporal variation in water velocity and generally low water velocities (relative to maximum velocities in upland streams) may result in the riverbed comprising mainly fine sand, silt and clay. The amount of sediment accumulating in the downstream reaches will also be determined by the amount of material being transported out of headwater reaches. Increased depth in large lowland rivers can also result in little light reaching the riverbed, particularly if the water is relatively turbid due to suspended silt or algae.

Potential habitats for living organisms also exist in the open water above the substrate, and in the interstitial water that flows through the substrates that make up the streambed. In small fast-flowing upland streams, a permanent open-water (pelagic) fauna seldom exists. Many organisms do however enter the water column and drift downstream for short periods (see Chapter 6). Such drifting behaviour typically exhibits a regular pattern of diel variation with drift densities usually being highest at night. Along larger or slower flowing lowland river reaches, true plankton communities may develop, similar in many ways to those that form in lakes.

5.8 What lives in running waters?

5.8.1 Plants

The presence of plants in the channels of rivers and streams is highly variable. In the forested headwaters of many streams, dense shading often prevents much growth of any aquatic plants within the channel, although bryophytes (mosses) may be abundant if the substrate is relatively stable. Turbulent flow and unstable substrates can also restrict plants severely. Growth of plants in the lowland reaches may be constrained by the combined effects of water depth and high turbidity that can reduce the amount of light available for photosynthesis on the riverbed. In such systems, the growth of plants (including benthic algae) may be restricted to the margins. Extensive beds of emergent reeds or rushes may flourish on the inside of river bends where banks are protected from the direct force of the water. The growth forms of macrophytes in running waters is similar to that seen in lakes (see Chapter 4, Fig. 4.4), although more fragile forms are generally restricted to slower flowing areas.

As mentioned previously, a stream cannot be easily separated from its catchment. Although not strictly growing within the stream, riparian vegetation is very much part of the stream ecosystem (Fig. 5.6). The roots of riparian trees such as willows (*Salix* spp.) frequently extend into the stream. Such roots can provide a stable substrate on which organisms live, and protect soft substrates from the erosive force of the flowing water. Overhanging vegetation can also determine the amount of light reaching the channel, and hence has a major influence on the rate of primary production occurring within the stream. Leaves, bark and wood that drop into the stream from overhanging vegetation are also a potential food source for heterotrophic organisms.

5.8.2 Microbial organisms

As in virtually all ecosystems, freshwater or otherwise, microbial organisms play a key role in primary production, breakdown and recycling of materials and nutrients in lotic systems. Touch the surface of any rock that has been in a stream for a few days and it will be slightly slimy.

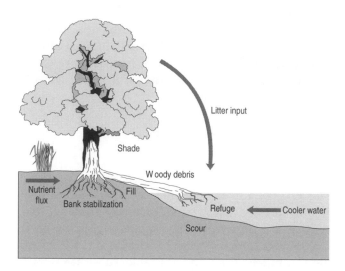

Fig. 5.6 Some of the roles that streamside (riparian) plants play in maintaining stream and river ecosystems. Shade cools the water, while fallen logs provide refuge for fish and help scour pools.

Slimy organic films (biofilms) grow on any hard surface that is present within a stream and consist of a complex assemblage of algae (particularly diatoms), bacteria, fungi and fine organic and inorganic particles (Fig. 5.7). The composition of biofilms will vary as environmental conditions change. In well-lit environments, the proportion of algae growing within a biofilm is higher whereas in poorly lit forest streams, the biofilm contains a higher proportion of bacteria and fungi. Varying amounts of inorganic and organic particles will also be trapped and incorporated into a biofilm. How much is trapped will depend on water

Fig. 5.7 A scanning electron micrograph of biofilm growing on woody debris in a New Zealand forest stream. (Photograph courtesy of Ross Thompson.)

velocity, the amount of particulate material being transported along the stream, the types of organisms growing within the biofilm and the degree to which various heterotrophic organisms are grazing and removing the biofilm.

As in lakes, microbial organisms play a key role in breaking down organic material that either falls into a stream (**allochthonous material**), or grows and dies within a stream (**autochthonous material**). This process of decomposition recycles organic carbon and other nutrients making them available to the stream fauna. In forested streams, little primary production occurs due to low light conditions. Consequently, the only organic matter available to heterotrophic organisms is likely to be derived from material that falls into the stream from riparian vegetation. The food quality of dead leaves and bark when it first falls into a stream is generally poor, consisting of little more than indigestible cellulose. However, colonization of such material by microbes markedly increases the content of proteins and complex carbohydrates within the decaying material. Many stream invertebrates that feed upon decaying leaves will preferentially feed upon leaves that have been in the stream for a few days rather than fresh leaf material. Invertebrates consuming decaying leaf material digest the colonizing microbes, their faeces consisting of little more than the remaining undigestable cellulose. These small particles of cellulose that have been through the invertebrate's gut can once again be colonized by microbes and may be consumed by detritivorous invertebrates that feed upon fine particulate organic matter.

True phytoplankton is usually only present in the lowland reaches of large, long rivers where shading from riparian vegetation is limited and the time taken for the water to reach the sea is measured in weeks to months. In the long, slow-flowing rivers, such as the Murray-Darling river system in eastern Australia, a combination of warm temperatures and high nutrient concentrations can result in very high densities of phytoplankton and associated declines in water quality (Hotzel & Croome 1994).

5.8.3 Invertebrates

Invertebrate communities in upland streams are dominated by various insect larvae and nymphs (Merritt & Cummins 1996), although nematodes, oligochaetes, gastropod and bivalve molluscs, and various crustaceans are also common. The larvae of case-building and free-living caddisflies (Trichoptera), true flies (Diptera) and beetles (Coleoptera) are usually diverse and abundant. These insects occupy a wide variety of niches with different species filtering fine particles from water, scraping biofilms from rocks and wood, consuming rotting leaves and other organic material, and burrowing through deposits of fine inorganic and organic sediments. Many free-living caddisfly and coleopteran larvae are highly efficient predators. Nymphs of mayflies (Ephemeroptera) and stoneflies (Plecoptera) also tend to be very abundant and diverse in well-oxygenated, cool streams. Again, different species pursue a range of lifestyles including grazing biofilms, consuming decaying organic matter and preying upon other invertebrates (see Section 5.9).

In contrast with lakes, crustaceans are rarely as abundant in small streams, although amphipods and isopods may be common. Various small species of harpacticoid and cyclopoid copepods may also be found living amongst various stream substrates. Omnivorous crayfish and shrimps may also be present in pools and backwaters and although not usually present in huge numbers, can have quite significant impacts on stream communities due to their relatively large size and capacity to consume large amounts of detritus (Usio & Townsend 2002). In large lowland rivers, a true zooplankton dominated by small crustaceans such as cladocerans and copepods may occur, and here there are parallels with lakes and other standing waters.

5.8.4 Vertebrates

As in lakes, fish are usually the most common and conspicuous vertebrate animals inhabiting river systems (Matthews 1998). Whereas many lakes lack fish due to their isolation from other aquatic systems, most rivers contain fish assemblages although fish-free sections of stream may occur upstream of barriers that block the upstream migration of fish. Small intermittent streams may also lack fish due to periodic drying. The constant presence of fish and the threat of predation they pose may be a major reason for the cryptic (small and well-camouflaged) nature of many stream animals (see Chapter 9).

Longitudinal zonation along river systems can also be seen in fish communities. Fish communities in small, upland streams tend to be relatively simple, often being dominated by one or two small generalist species with wide physiological tolerances. Such fish may have to tolerate periods of severe cold or turbulent floods, or the opposite extreme of high temperature and low flow during droughts. Fish community diversity generally increases moving downstream. The stability and range of habitats and the opportunities for specialization increase, although degradation of downstream habitats due to pollution and other forms of habitat disruption can severely disrupt such patterns.

Various mammals, such as the platypus and white-tailed water rat in Australia and various species of otter across Eurasia and North and South America, can also act as significant predators on stream fauna. Beavers are important 'ecosystem engineers' in some streams in the northern temperate forests of North America through their construction of dams across channels.

Some birds are river and stream specialists. The dipper of Europe and the blue duck or whio of New Zealand specialize in feeding on invertebrates picked off rocks in fast-flowing streams. Although not truly aquatic, numerous insectivorous birds and spiders live along rivers and streams and feed upon the huge numbers of adult insects that hatch from rivers and streams. The impact that such terrestrial predators have on stream communities is largely unknown but may not be quite as high as shown by isotope studies (Sanzone *et al.* 2003).

5.9 Food webs in streams and rivers

In small streams, the organic carbon that forms the base of the food web comes primarily from either instream benthic primary production or falls into the stream in the form of leaves, bark and wood from terrestrial vegetation (Fig. 5.8). Wood and leaf material from terrestrial plants, often collectively described as **coarse particulate organic matter** (CPOM), is consumed by various invertebrates known as shredders. Shredders typically have powerful jaws that allow them to break large pieces of detritus into smaller pieces that can be consumed (Cummins *et al.* 1989). As mentioned in Section 5.8.2, CPOM is preferentially consumed after it has been colonized and partially broken down by a diverse assemblage of microbes. The microbes significantly enhance the nutritive value of the CPOM.

As in lakes, the biofilms that develop on any hard surface in streams are consumed by a group of invertebrates collectively known as grazers

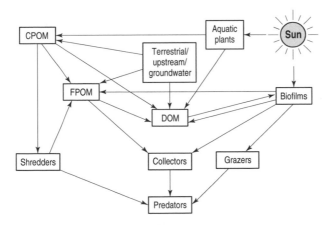

Fig. 5.8 Key energy pathways in rivers and streams. The importance of the various pathways is likely to vary greatly across different systems. Pathways leading from various consumers to CPOM, FPOM and DOM detrital groups have been omitted for simplicity. CPOM and FPOM categories also include various microbial organisms which colonize organic detritus and form an important food resource for consumers.

(Fig. 5.8). In well-lit areas, algae growing within the biofilm may be a key source of organic carbon for the stream food web. In poorly lit areas where bacteria and fungi are likely to dominate the biofilm, **dissolved organic carbon** (DOC) leaching from CPOM and **fine particulate organic matter** (FPOM) and washing into the stream from terrestrial sources, provide the energy that the heterotrophic bacteria and fungi require.

The faeces that both shredders and grazers release into the stream is a major source of FPOM. Shredders, in particular, play a key role in converting CPOM to FPOM (Fig. 5.8). As described in Section 5.8.2, various microbial organisms colonize FPOM, thus enhancing its nutritive value. FPOM accumulates in slow-flowing areas, such as the many spaces that occur between cobbles and in slow-flowing areas such as pools. Invertebrates that consume FPOM are collectively known as collectors. Those that consume the material from the bed of the river or stream are frequently described as collector–gatherers, whilst those that filter FPOM from the water column are known as collector–filterers. Switches from one feeding group category to another are common. Many shredders start life as collectors, only switching to feeding on CPOM as they become larger and their jaws more powerful (Cummins 1974).

Shredders, grazers and collectors may all be consumed by the wide variety of predators, both invertebrate and vertebrate, that inhabit streams. Switches from a detritivorous lifestyle to a predatory lifestyle are also common. Many invertebrate predators start life as collectors, feeding on FPOM, and only switch to a predatory lifestyle once they are large enough to catch and handle other invertebrates. Predatory fish also exhibit major shifts in their preferred food over their lifecycle. Many fish initially start feeding on organisms such as protozoa and rotifers, then progressively shift to feeding on larger invertebrates, and finally fish as they mature.

The relative importance of different resources and the abundance of the organisms dependent on them is strongly influenced by the surrounding catchment, the nature of the riparian vegetation and the size of the watercourse. A conceptual model, known as the river continuum concept (RCC), of how these various factors might interact over an entire river system has been developed (Vannote *et al.* 1980). The RCC argues that the proportions of shredders, collectors and grazers within a stream community will alter depending on the size of the river channel and the relative availability of instream primary production, CPOM and FPOM (Fig. 5.9). In small, shaded upland streams, large inputs of CPOM provide an abundance of food for shredders. Heavy shading however, limits the productivity of biofilms hence grazers are scarce. Further downstream, increasing width allows greater light penetration thus allowing greater biofilm productivity and increases in the abundance of

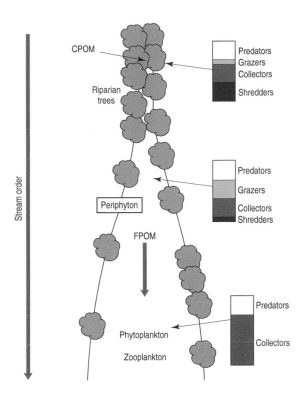

Fig. 5.9 The river continuum concept predicts that the proportion of invertebrates with particular feeding strategies will alter predictably with respect to stream size. (After Vannote *et al.* 1980 with permission of the National Research Council of Canada.)

grazers. CPOM will be less abundant but FPOM washed downstream may support an abundant community of filtering and gathering collectors. In large lowland rivers, limited light penetration can reduce primary productivity, and the community may be heavily dependent on supplies of FPOM washing downstream.

The value of the RCC as a useful description of stream ecosystem function has been much debated and challenged since it was first proposed (for excellent summaries of aspects of this debate, see Winterbourn *et al.* 1981, Gomi *et al.* 2002). Clearly, changes in the structure of the riparian vegetation, particularly in the headwater reaches, have the potential to alter greatly the relative importance of CPOM inputs and benthic primary production. Remove the riparian vegetation and we might expect CPOM inputs to fall, and benthic primary production to increase due to the increased light inputs. Alternatively, the construction of dams across rivers blocks the upstream–downstream transport of FPOM, and hence densities of collectors might be expected to be low immediately downstream of a major impoundment. Such ideas are hypotheses we can test by collecting data and conducting experiments, e.g. adding FPOM or CPOM and monitoring for community responses. The RCC highlights the importance of having a useful conceptual model from which hypotheses can be developed and then tested. There are other models, such as the riverine productivity model by Thorp and Delong (1994), so we may need to choose the model that is most appropriate for our purposes.

5.10   What to study in streams and rivers?

As discussed in Chapter 4 on lakes, a knowledge of the common organisms and the key physical and chemical features of a habitat allow us to start making more meaningful observations, noticing patterns and asking questions as to why a particular organism lives where it does. This is what ecology is all about. In rivers and streams, watching patterns of flow along a river and imagining how variation in flow might create different habitats for various animals and plants is a good starting point. A huge range of sampling tools for different situations is available for sampling communities in rivers and streams (Merritt *et al.* 1996). However, simply collecting and observing organisms are only the starting points of a potential scientific investigation. We then must quantitatively measure the pattern, develop conceptual models that can account for the occurrence of the pattern, propose falsifiable hypotheses that allow us to test the value of our models, and finally devise and complete the empirical tests of the hypotheses themselves (see Chapter 1). There is invariably no shortage of interesting patterns to observe and study in even the smallest of stream systems. It is only our powers of observation and imagination that limit the opportunities for study and further research.

5.11   Further reading

Cummins K.W. (1974) Structure and function of stream ecosystems. *BioScience* **24**, 631–641.
Cummins K.W., Wilzbach M.A., Gates D.M., Perry J.B. & Taliaferro W.B. (1989) Shredders and riparian vegetation. *BioScience* **39**, 24–30.
Gomi T., Sidle R.C. & Richardson J.S. (2002) Understanding processes and downstream linkages of headwater systems. *BioScience* **52**, 905–916.
Hotzel G. & Croome R. (1994) Long-term phytoplankton monitoring of the Darling River at Burtundy, New South Wales: Incidence and significance of cyanobacterial blooms. *Australian Journal of Marine and Freshwater Research* **45**, 747–759.
Matthews W.J. (1998) *Patterns in Freshwater Fish Ecology*. Chapman & Hall, New York.

Merritt R.W. & Cummins K.W. (1996) *An Introduction to the Aquatic Insects of North America*. Kendall/Hunt Publishing, Dubuque, Iowa.

Sanzone D.M., Meyer J.L., Marti E., Gardiner E.P., Tank J.L. & Grimm N.B. (2003) Carbon and nitrogen transfer from a desert stream to riparian predators. *Oecologia* **134**, 238–250.

Thorp J.H. & Delong M.D. (1994) The riverine productivity model—an heuristic view of carbon-sources and organic-processing in large river ecosystems. *Oikos* **70**, 305–308.

Usio N. & Townsend C.R. (2002) Functional significance of crayfish in stream food webs: roles of ominivory, substrate heterogeneity and sex. *Oikos* **98**, 512–522.

Vannote R.L., Minshall G.W., Cummins K.W., Sedell J.R. & Cushing C.E. (1980) The river continuum concept. *Canadian Journal of Fisheries and Aquatic Sciences* **37**, 130–137.

Williams D.D. & Feltmate B.W. (1992) *Aquatic Insects*. CAB International, Wallingford.

Winterbourn M.J., Rounick J.S. & Cowie B. (1981) Are New Zealand stream ecosystems really different? *New Zealand Journal of Marine and Freshwater Research* **15**, 321–328.

Part 2   *Fundamental Ecological Questions*

In the first part of this book, we discuss some of the intellectual and practical tools freshwater ecologists use to understand the factors and processes that determine patterns in nature. In the next two parts, we review how we can use these tools to explore key questions in freshwater ecological research. Each of these chapters begins with an observation of an ecological phenomenon. That observation provides a starting point from which we can develop conceptual models that describe how we think a particular system might work – the 'scientific method' in action! From there, we explore diverse approaches that researchers have used to address the various hypotheses and questions that can be derived from good models.

In Part 2, we examine four **fundamental ecological questions** that aquatic ecologists ask when assessing the distribution and abundance of organisms in freshwater systems. We use the term fundamental ecological questions to identify key ecological processes that can play a role in determining the abundance of organisms in any ecosystem, freshwater or otherwise. Understanding how dispersal, habitat selection, disturbance and predation structure ecosystems is of intrinsic scientific interest, quite apart from any applications such knowledge might have.

When reading each chapter, it is important to think critically about the approach used by various researchers to address each question and test hypotheses. The various examples chosen represent researchers' best attempts at addressing the various questions. Keep in mind the constraints they are likely to be working under, including limits to the resources available to undertake the study, challenges of the various habitats in which they are working, difficulties associated with sampling at small or large spatial or temporal scales or perhaps gaps in the knowledge of how a particular system works at the time the study was undertaken. A phenomenon that is well understood and seemingly obvious today may only seem that way due to the hard work that some researcher has completed previously. However, all science is open to challenge. As you read the various studies, ask yourself: Did they get it right? Could you do better? How?

Chapter 6   *How are population numbers and structure affected by dispersal?*

6.1   It rains – and previously dry beds gradually boil with life

A drought grips western Victoria in south-eastern Australia. Such events are not unusual in Australia, a country with such large variability in rainfall that droughts and floods are frequent events. A small pond, no more than 15 m across has completely dried up, its existence as an occasional waterbody marked only by a shallow depression in the ground and the presence of sedges around its boundaries. Finally in mid-winter, heavy rains fall. The pond begins to fill, and within a matter of days small animals can be found flitting through the water column. They are tiny planktonic crustaceans called *Boeckella* and *Mesocyclops*, and they are feeding on tiny particles in the water column: small planktonic algae and even bacteria. Initial species of such grazers disappear to be replaced by others with similar resource requirements. Gradually the number of predatory species rises. Altogether, some 90 species come and go in a seemingly chaotic and unpredictable way before the pond dries again several months later.

A short distance away, the drought has also dried the Lerderderg River to a series of puddles in the deeper sections of the bed. Soon after flow ceased and the riffles dried, species richness rose in the pools apparently due to some emigration from the riffles. Densities rose as the pools dried, and the fauna became dominated by predaceous invertebrates. Gradually the stream-dwelling invertebrates disappeared and the late summer pool fauna comprised typically lentic taxa such as dytiscid beetles, chironomid midges and case-building caddisflies. Finally, even the puddles dried. The beetles flew away or burrowed into the leaf litter at the bottom of the pool. The caddisflies and midges either emerged in time as adults or perished. After a few months, flow resumed. Some of the initial 'pioneer' taxa were species that apparently had desiccation-resistant eggs. Some midge larvae of a different group from the summer pool genus were also early colonists. If pools had persisted upstream, case-building caddisflies and mayflies were washed down into the newly wetted bed. After a few weeks, the pioneer taxa were less common and the fauna was now more typical of nearby permanent streams. Many other species had colonized, presumably flying in from permanent streams or pools a short distance away. A few species were tolerant of drying or persisted under the leaf litter on the streambed and in the damp sediments below. However, they were not immediately abundant when flow began, as if to guard against 'false starts'. As was seen in the temporary pond, the sequence of arrivals and identities of species that gained high abundance seemed driven greatly by chance dispersal events.

The above, true stories are dramatic examples of how the numbers of organisms present at a locale can be determined greatly by dispersal. Dispersal is a ubiquitous aspect of natural systems – ongoing even in environments not subject to such dramatic changes as temporary ponds and rivers. So, what role can dispersal play in ultimately determining the population numbers of species?

6.2   Some basic population ecology

Before we can begin to answer the above question, we need to cover some basic principles of population biology, and talk about what drives population growth. A **population** is defined as all the individuals of a species living within a defined area of space. It is important that all population members are capable of interbreeding with each other, which is why we specify that they all be members of the same species. Additionally, individuals of the same species are not members of the same population if they are sufficiently separated in space that they never interbreed.

Populations can increase in number in two ways: via the production of new individuals through reproduction (usually referred to as **births**) and by individuals dispersing into the population from elsewhere (**immigrants**). Likewise, populations can decrease in size in two ways: through the **death** of individuals and by **emigration**, that is, individuals dispersing away from the population. All of these four processes are usually expressed as *rates*, that is as the numbers (of births, deaths, immigrants or emigrants) per unit time. These concepts of population apply to solitary, rather than colonial, organisms (refer back to Box 2.1) because it is difficult to speak of the birth and death of individuals when they live together intimately and share resources as part of one colony.

How are population numbers and structure affected by dispersal?

Populations grow when the numbers (or rates, if we express them per unit time) of births and immigrants collectively exceed those of deaths and emigrants. Likewise, the population will shrink in size if the rates of deaths and emigration collectively exceed that of births and immigration. We define **population growth rate** as the per capita average rate of increase of population numbers, and it is often symbolized by the letter r (Fig. 6.1). If the population is actually declining in size then population growth is simply expressed as a negative number, strange as this sounds. The other concept we should introduce briefly is that of **carrying capacity**, which is the maximum number of individuals that can be sustainably supported by the amount of food, living space and other resources that are available – this is usually symbolized by K (Fig. 6.1). Carrying capacities of populations are usually difficult to measure exactly because they are likely to vary with time (for example, because amounts of food usually fluctuate) and hence are often discussed in approximate, rather than exact, terms.

The concept of carrying capacity is useful in demonstrating that populations cannot increase indefinitely. As population numbers exceed K, individuals will begin to run out of things like food and territory, and this causes many deaths or much emigration and a plunge in numbers. Likewise, numbers cannot decline below zero – a population that reaches zero has gone extinct, and can only be re-established by immigrants from elsewhere. The possibility that populations go extinct and are renewed is captured in the concept of **metapopulations**

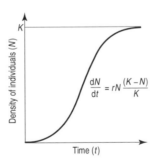

$$\frac{dN}{dt} = rN\frac{(K-N)}{K}$$

Fig. 6.1 Density of individuals (N) increasing in a population over time (t). Population growth is often (for continuously breeding species) modelled by an S-shaped (or sigmoidal) curve like this one. The equation for this line, given in the figure, is called the logistic equation. The change in N for a given change in time (dN/dt) depends upon the population growth rate, r (the slope of the line), and how close the value of N is to K, the carrying capacity. As N approaches K, then $(K - N)/K$ becomes a very small number, meaning that the line (and population numbers) levels off at K. The logistic equation is a useful way of modelling populations, but population densities rarely show such smooth increases and levelling out in nature.

(Fig. 6.2). Metapopulations are groups of populations. The populations are separated in space and are connected by dispersal of individuals between them. Some of these populations may be **sources** where births exceed deaths, and others may be **sinks** where deaths exceed births. Source populations can support sink populations through dispersal of individuals, and it is the frequency of dispersal between populations that determines the structure of the metapopulation (Fig. 6.2). Because dispersal in nature is quite difficult to measure (especially for small or cryptic animals and for tiny seeds or pollen), it is unclear how many populations are part of a 'classic' metapopulation structure (Fig. 6.2a) as opposed to being arranged in one of the other combinations. In particular, we usually do not know whether we have one big, patchily distributed population (Fig. 6.2c) or a series of populations, which may or may not (Fig. 6.2a,b,d,e) be connected by dispersal.

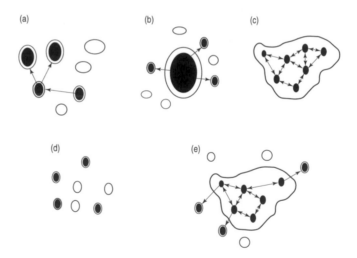

Fig. 6.2 Different types of metapopulations. The filled symbols indicate habitat patches occupied by individuals, and the unfilled symbols indicate vacant patches of habitat. The arrows indicate dispersal. The outer, black lines that encompass filled symbols are the boundaries of populations. (a) The classical metapopulation, in which populations have an equal chance of going extinct and overall persistence of the metapopulation relies on recolonization of habitat patches from source populations. (b) So-called 'mainland-island' dynamics, in which one (or more) populations has larger size or greater longevity and acts as a source population for other, smaller, less long-lived populations. (c) A patchy population, in which dispersal is sufficiently frequent between habitat patches that all individuals are actually part of the same population. (d) Populations where dispersal of individuals between them is rare or non-existent, so that a habitat patch where the population goes extinct is not recolonized. (e) A case that combines features from (a–d). (After Harrison & Hastings 1996 with permission from Elsevier Science.)

6.2.1 A big question

Zero and the carrying capacity, K, are the absolute bounds within which a population can persist sustainably. A population that has reached zero (or where the remaining individuals cannot reproduce, such as an all-male population in a species that requires sexual reproduction) has gone extinct. When a population exceeds K, and hence exceeds the limits of food and other resources, high mortality or emigration rates will cause a crash. Nevertheless, many populations do not seem to approach either of these extremes very often. It may be that many populations are actually regulated around some intermediate abundance, such that positive growth happens when densities are low and negative growth when densities are high. Whether and how such regulation occurs are central questions in basic population ecology.

Either way, it is clear that population numbers can be constrained in different ways in different species: by low birth rates, high mortality rates or high emigration rates. Moreover, many organisms have complex life-histories involving multiple stages, some of which may live in quite disparate environments. If we are interested in questions about population numbers, we must remember that limits could be set anywhere during the lifecycle. For example, population numbers of some macro-invertebrates could be determined largely by high mortality rates among adults, but we are unlikely to realize this if we focus our attention solely on the juvenile stages.

6.3 Can dispersal limit sizes and occurrences of freshwater populations?

You should be able to see now why the dispersal of individuals into and out of populations can be very important. Immigration and emigration, as basic population processes, can potentially set limits to population numbers (along with, of course, birth and mortality rates). Additionally, dispersal between populations is important in a metapopulation context, because some populations may have to have relatively regular immigration if they are to persist. Dispersal is especially critical where a habitat must be periodically recolonized anew because conditions become harsh or uninhabitable (e.g. droughts and floods). Dispersal, however, is an ongoing feature of all environments, including those not predisposed to such large fluctuations in habitability. So, can populations be limited by dispersal rates and if so, under what sorts of conditions? We now return to consider the dispersal methods used by freshwater organisms and whether dispersal frequencies may limit population numbers or affect population structure in freshwater habitats.

6.3.1 Dispersal methods in freshwater organisms

Freshwater organisms have a number of ways of moving about: they can swim through water, they can walk along the bottom or the water's surface and, because some of them live on top of the water or have terrestrial stages, some can fly. When thinking about their dispersal capabilities, we should keep a couple of issues in mind. First, water currents or wind can disperse small organisms over distances that easily exceed those they could travel under their own power by swimming, walking or flying. In lakes and ponds, there are no ubiquitous currents, of course. Nevertheless, prevailing winds can whip up waves and create currents, especially on larger waterbodies, that will transport many organisms to leeward margins.

Water flow is probably responsible for a lot of distance covered by micro- and macroinvertebrates, including both benthic and planktonic forms. In streams and rivers, plants are often dispersed by water flow, which can move seeds, spores, stems, and other plant parts capable of vegetative reproduction to downstream locations. The transport of organisms downstream by water flow is termed **drift** (see Section 5.7), a form of dispersal. **Passive dispersal** is when organisms or plant parts are dispersed entirely by water currents, simply dropping to the bottom when the flow is no longer able to buoy them up. However, many invertebrates can change their dispersal distance markedly simply by swimming either up or down, which can expose them to different water velocities (faster and slower, respectively). Some organisms, like the larvae of blackflies, play out silk lines, which then snag on objects and allow them to pull themselves back down to the bottom. When invertebrates make such choices, this is termed **active dispersal**, because the organism no longer behaves as if it were an inert particle with no control over where it goes.

However, even when organisms are dispersed passively, our ability to predict where they will end up can still be quite poor (we will talk about this more in Chapter 7). Water velocities will, on average, increase with distance from the substratum, but the complexity of most river beds (which are often covered by boulders or gravel) means that velocities just above the bed are hugely variable. It is possible to find spots even in fast-flowing areas of streams where the water is stationary or even travelling upstream, the latter often occurring in dead water zones immediately downstream of large boulders or logs. Dead water zones are often small in volume of course, but we should keep in mind that many lotic fauna and flora are also small, so these zones may be critical to them as places to get out of the current.

The same sorts of issues apply to aerial forms. Wind can often push flying invertebrates in particular directions, and the same considerations in terms of active and passive dispersal apply, as well as the virtual unpredictability of wind speed and direction within vegetation or close to the ground. Occasionally, plants and animals are caught up in wind spouts and transported great distances, some of which might result in successful dispersal into another waterbody. Perhaps only the larger invertebrates (e.g. dragonflies) and fish species, birds and mammals are capable of exceeding water and wind currents to travel almost always in the directions of their choosing. The upstream migration of salmon from the sea to the headwaters of rivers is a particularly impressive example of this. Some watermites attach themselves to dragonflies and beetles and are transported between waterbodies. Waterfowl can also disperse invertebrates, seeds and spores in the mud that sticks to their feet.

Freshwater organisms, then, have a good array of dispersal routes that are potentially open to them. Dispersal can occur during both the juvenile and adult stages, and for some taxa this means dispersal through both water and air. If we are interested in whether dispersal affects population numbers, we need to study more than one life-history stage.

6.4 Two models of how dispersal might affect population numbers

Let us consider two models of dispersal, each of which might be considered to be at one end of a continuum of possibilities for a population. In the first model (Model 1), successful dispersal – that is, dispersal where individuals survive dispersal and immigrate successfully – is a frequent event. There is often a saturation of numbers at any locality and rarely any shortage of potential colonists when habitat becomes available. In this situation, we may have a patchy population, with lots of dispersal among patches (Fig. 6.2c). In the second model (Model 2), at the other end of the continuum, successful dispersal is much less frequent. Perhaps many dispersing individuals are swept into unsuitable habitats, starve or are eaten by predators before they manage to settle into a population. Occasionally, a chance combination of factors may allow many of the dispersing individuals to survive, leading to larger numbers of successful immigrants and large fluctuations in time and space in the influence of dispersal. In this model, we can see that populations are sometimes constrained by immigration, and the most abundant species present may be those fortunate to have many dispersing individuals present at the right time. This type of situation may be particularly pertinent for species where juvenile stages end up dispersing long distances from where they hatched. In such cases, populations may rely on waves of immigrating juveniles to provide the next generation of adults, and we may have a metapopulation structure where some populations are supported by others by occasional dispersal (Fig. 6.21a, b or e). Alternatively, when dispersal is always unsuccessful, we may have a series of completely isolated populations (Fig. 6.2d).

So, what is the evidence that dispersal events shape the numbers and identities of organisms present compared with the possibility that immigrants are always in excess and hence do not constrain numbers? Of the types of population structures illustrated in Fig. 6.2, which best describes freshwater species? We will consider some interesting research results from both streams and temporary ponds to examine these questions.

6.5   Dispersal and colonization by lotic organisms

It is probably fair to say that stream researchers looking at macroinvertebrates initially believed that dispersal was frequent, with individuals successfully covering quite large distances. Many researchers had observed that macroinvertebrate drift was a ubiquitous event in streams, occurring mostly at night, and occasionally comprising high densities of organisms (see Section 5.7). These observations led Müller to propose a model called the colonization cycle, in which larvae of species having terrestrial adults disperse long distances downstream and adults fly back upstream to colonize supposedly denuded upstream sections (Fig. 6.3). This model proposes that groups of individuals are well mixed along stream sections by frequent and successful dispersal during both larval and adult stages. This initial model was thus more like Model 1 than Model 2 (although researchers did not put it into that population context). We would expect that individuals in the same stream channel would comprise one big patchy population (as different parts are connected by dispersal) and, if dispersal is as frequent as observations of drift had suggested, then numbers of individuals at any one location should not often be limited by unsuccessful dispersal. (Note that the colonization cycle model has been criticized for using 'group selection' to explain adult behaviour. We will not go into this here because our focus is on dispersal and its implications, but interested readers should consult Anholt (1995) for more information.)

Is the evidence consistent with this view of streams and rivers? Although the colonization cycle is an appealing model given the ubiquitous nature of drift, dispersal by individual invertebrates is difficult to measure. We want to know how far and how frequently each individual travels without dying before reaching the bottom again, but simple measurements of the density of animals in the drift at particular points in time and space do not provide that information. It is possible that

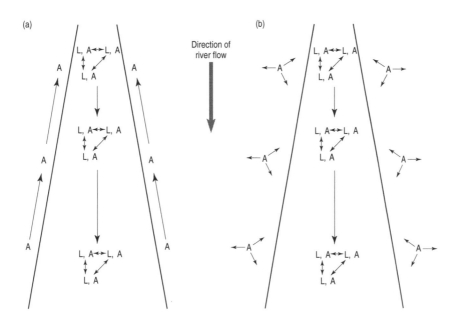

Fig. 6.3 Schematic diagrams illustrating two (of many) possibilities for dispersal in and among lotic populations. The two lines indicate a length of river. L indicates larval or nymphal stages and A indicates adult stages, with some taxa having fully aquatic life-histories and other taxa (most of the insects) having terrestrial, winged adults. The arrows indicate dispersal with short arrows indicating small, within-habitat movements and long arrows dispersal between habitat patches. (a) The length of river is inhabited by one population connected by dispersal up- and downstream. Larvae and aquatic adults move in all directions within habitat patches and can disperse in a downstream direction using the current. Adults of species with terrestrial winged forms fly upstream to recolonize upstream sections in a model usually known as the colonization cycle. (b) The length of river is inhabited by a series of separate isolated populations. Juveniles and aquatic adults move in all directions within habitat patches but there is no dispersal to downstream sections using flow and adults do not disperse upstream, downstream or overland.

many animals are only travelling very small distances – more akin to small foraging movements than to true dispersal – or that many drifters suffer relatively high mortality, either because they drift as a last resort to find food or are picked off by fish. Some evidence for both of these latter suggestions has been found, leaving the prospect that drifting is not necessarily a successful method of dispersal after all.

The other side of the coin concerns adults, which should disperse away from the point they exit the stream and fly predominantly upstream to recolonize headwater sections. Measuring adult dispersal is as difficult as measuring larval dispersal, but we can use traps that distinguish which direction adults were travelling when trapped. Some studies showed the expected higher numbers of adults travelling upstream compared to downstream, but other studies have shown adults fly in all directions including sideways into the riparian vegetation or directions controlled most by prevailing winds. The evidence for intense waves of upstream dispersing adults is not convincing.

Nevertheless, the real difficulty is that most of the available data cannot definitely rule out one alternative dispersal model over another. It is only relatively recently that some researchers have begun to collect more compelling and unambiguous evidence. We shall discuss two studies, one that supports the colonization cycle as a good description of both population boundaries and of dispersal, and one that suggests the opposite. Such results suggest that there will not be one universal model of dispersal and population structure to describe all aquatic invertebrates. This is not entirely surprising, and the challenge will be to discover how and why stream taxa differ from each other.

6.5.1 Dispersal by the mayfly *Baetis* in an Arctic river

Our first study is by Hershey *et al.* (1993), who looked at dispersal by the mayfly *Baetis* in the Kuparuk River, arctic Alaska. This research group has been studying how adding extra phosphorus to the low nutrient river water boosts algal growth and changes trophic interaction within the fairly simple macroinvertebrate community present. *Baetis*, an algal grazer, increased in density in the fertilized section of the river, probably because of the increased amounts of food available. This increased density had to be due to drift of *Baetis* from upstream sections and not due to small instars hatching from locally laid eggs after nutrient additions had begun. *Baetis* does not have multiple generations per year in this river, and egg-laying and hatching had already occurred prior to nutrient addition.

Hershey *et al.* (1993) reasoned that they could use this situation to test whether the colonization cycle applies to *Baetis*. They compared drift densities at multiple locations both upstream and downstream of the fertilized section over 11 different dates, and showed that drift densities were typically higher above the fertilized section than downstream of it. They looked at the density of *Baetis* and showed that early in the season, when most *Baetis* were small hatchlings, there were no systematic differences in benthic densities between upstream and downstream sections. Later in the season, benthic densities were typically much higher downstream than upstream. All these observations on drift and benthic density are consistent with the model that *Baetis* nymphs drift downstream as they grow. Hershey *et al.* (1993) estimated that approximately one-third to one-half of the upstream juvenile population drifts downstream into the fertilized section each summer.

So, are the adults emerging from these downstream locations then flying back upstream to recolonize these sections? Hershey *et al.* (1993) dripped a stable isotope of nitrogen, ^{15}N, into the river to tag *Baetis* nymphs in the fertilized, downstream sections. (^{15}N is otherwise rare naturally, so the proportion of ^{15}N present in an insect's body is a good indication of exposure to the ^{15}N that has been added at one location in the river). After some time, they collected nymphal and adult *Baetis* both downstream and upstream of the ^{15}N addition site (Fig. 6.4). They found that adults caught downstream were indeed tagged with ^{15}N, which they had originally consumed as nymphs. Nevertheless, the isotopic signal strength was lower for these adults than it was for nymphs sampled at the same location. This suggests that adults comprised individuals that had emerged at locations with different isotopic signals – that is, at points of differing distance from the ^{15}N dripper. Hershey *et al.* (1993) also found adults tagged with ^{15}N at locations upstream of the ^{15}N dripper.

All of these observations are consistent with the notion that adults are flying away from their emergence sites and are going preferentially upstream. In fact, the best-fitting mathematical model applied to the data (Fig. 6.4) suggested that about one-third to one-half of adults were flying about 1.6–1.9 km upstream. The colonization cycle thus seems to be a reasonably good description of dispersal dynamics for this population of *Baetis*, which is probably well connected by dispersal up and down the channel.

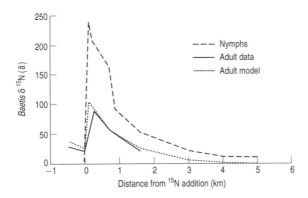

Fig. 6.4 Proportion of nymph and adult *Baetis* with ^{15}N signals upstream and downstream of the ^{15}N addition in the Kuparuk River and predictions from a mathematical model for the adults that presumes adults trapped at any one point are a mix of those that emerged at that point and adults that emerged downstream (where they were exposed to less ^{15}N) and flew upstream. Note that the mathematical model fits the adult data well. Note also that adults with a ^{15}N signal are found a significant distance upstream of the dripper, whereas the nymphs are not, consistent with the expectations of the colonization cycle. (After Hershey *et al.* 1993 with permission of the Ecological Society of America.)

6.5.2 Dispersal and the shrimp *Paratya* and the mayfly *Baetis* in tropical streams

We now turn to our second study, which was carried out in Queensland by Hughes *et al.* (1995) and which also used a clever, indirect method of measuring dispersal. The markers used in this study were genetic markers – the frequencies of particular alleles (which are different forms of genes occupying the same locus) in groups of individuals collected from different places along and between stream channels. If the frequencies of alleles are very different in different places, we can infer (given a certain set of assumptions that we shall not discuss here) that dispersal of individuals between these places is rare or non-existent. Hughes *et al.* (1995) looked at a freshwater shrimp called *Paratya* that, because it has a planktonic larval stage, was thought likely to disperse frequently along stream channels. However, individuals collected from different locations had very large differences in allele frequencies – in particular, there were large differences between individuals separated by only a few kilometres of river distance (Fig. 6.5).

It would seem that *Paratya*, despite its planktonic larval stage, either does not disperse large distances or suffers virtually 100% mortality before reaching new locations. Models like Model 1 do not describe its dispersal dynamics, and the underlying population structure may consist of a series of isolated, separate populations despite the connection by water flow. Bunn and Hughes (1997) have since reported similar studies on a number of insects, including an Australian species of *Baetis*. The allelic frequencies of these taxa suggested that, unlike *Paratya*, they disperse often between catchments and subcatchments and this was probably through aerial dispersal by adults. Nevertheless, there were lots of allelic differences between nymphs collected in different reaches but in the same stream. It looked as though larvae disperse only short distances and may have been generated from only a few successful matings. Far more species need to be examined, but these tantalizing results suggest that we have much to learn about both dispersal and population structure in stream organisms. Certainly the colonization cycle is not a good model of the dispersal dynamics of fauna in these tropical streams.

6.6 Dispersal and colonization in lentic organisms

How does dispersal by the organisms that live in lakes and ponds potentially determine population numbers and abundance? Drake (1991) was interested in the assemblage of organisms that commonly live in the water column of natural ponds and lakes in north-central

How are population numbers and structure affected by dispersal?

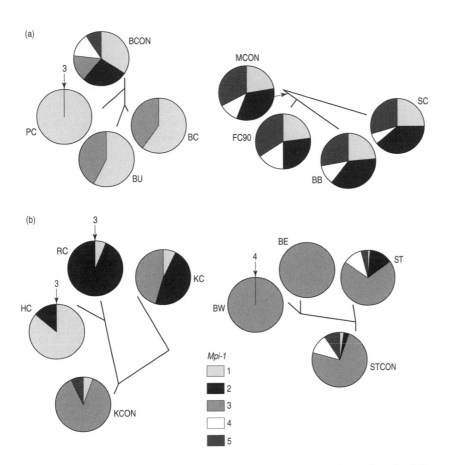

Fig. 6.5 Allelic frequencies at one locus (the *Mpi-1* locus) that were detected in the shrimp *Paratya australiensis*. There are two catchments (the Mary River (a) and the Brisbane River (b)), each with two subcatchments. *Paratya* were sampled from each of three upland sites in each subcatchment as well as one site in each subcatchment that was downstream of all the other three. The position of the pies in each subcatchment reflects the spatial arrangement of sites and the lines connecting them to the stream distance. Note the large differences in allelic frequencies between catchments and subcatchments, but also between sites within the same subcatchment and that were connected by water flow. BB, Broken Bridge Creek; BC, Boolumba Creek; BCON, Boolumba Creek confluence; BE, Branch Creek east; BU, Bundaroo Creek; BW, Branch Creek west; FC90, Flagstone Creek in 1990; HC, Humbug Creek; KC, Kilcoy Creek east; KCON, Kilcoy Creek confluence; MCON, upper Mary River confluence; PC, Peters Creek; RC, Rum Crossing; SC, Scrubby Creek; ST, Stony Creek; STCON, Stony Creek confluence. (After Hughes *et al.* 1995 with permission of JNABS.)

Indiana, USA. These organisms included several species of algae (such as the green algae *Chlamydomonas*), and several species of invertebrates, mostly crustaceans, that either grazed the algae or were predators upon other invertebrates.

Drake (1991) set up replicate aquaria and introduced 13 species into the aquaria using 10 different sequences of invasion. He had three replicates of each sequence and followed each aquarium for 195 days to see whether all aquaria ended up with the same array of species in the same abundance at the end of the experiment. What Drake found was that the sequence in which species were introduced into the aquaria had a profound effect upon the final identities and abundances of species at the end. In 40% of aquaria, the dominant alga at the end of the experiment was the species that was introduced first. Nevertheless, in some sequences the initial algal colonizer did not maintain numerical dominance. Outcomes for the grazing and predatory species were even more heterogeneous – some sequences proved resistant to subsequent invasion by consumer species, which quickly went extinct, whereas others were readily invaded.

The results led Drake (1991) to conclude that timing, chance and the sequence in which species arrive play strong roles in determining which species attain high population numbers. Of course, this is a laboratory experiment, but if the outcomes are this variable under relatively simplified and controlled conditions, it suggests that the recolonization of dried pond and riverbeds in nature might also be strongly affected by the chance arrival of taxa that happen to be dispersing at the right moment.

6.7   Conclusion

Dispersal plays a central role in both the structure of populations – that is, whether we have patchy or isolated populations or a metapopulation – and in determining the numbers within populations. Of course, birth and mortality rates are equally important to consider in any questions concerning population dynamics. By focusing this chapter on dispersal, we are not suggesting this process somehow has greater significance to understanding population numbers than studies of births and deaths. It is simply that recent research on dispersal is suggesting that our current views of populations of freshwater organisms could be profoundly wrong in many cases. It seems likely that such interesting findings will spur researchers into asking more directed questions about population regulation in freshwater organisms.

6.8   Further reading

Anholt B.R. (1995) Density-dependence resolves the stream drift paradox. *Ecology* **76**, 2235–2239.

Boulton A.J. (1989) Over-summering refuges of aquatic macroinvertebrates in two intermittent streams in central Victoria. *Transactions of the Royal Society of South Australia* **113**, 23–34.

Boulton A.J. & Lake P.S. (1990) The ecology of two intermittent streams in Victoria, Australia. I. Multivariate analyses of physicochemical features. *Freshwater Biology* **24**, 123–141.

Boulton A.J. & Lake P.S. (1992a) The ecology of two intermittent streams in Victoria, Australia. II. Comparisons of faunal composition between habitats, rivers and years. *Freshwater Biology* **27**, 99–121.

Boulton A.J. & Lake P.S. (1992b) The ecology of two intermittent streams in Victoria, Australia. III. Temporal changes in faunal composition. *Freshwater Biology* **27**, 123–138.

Bunn S.E. & Hughes J.M. (1997) Dispersal and recruitment in streams: evidence from genetic studies. *Journal of the North American Benthological Society* **16**, 338–346.

Downes B.J. & Keough M.J. (1998) Scaling of colonization processes in streams: parallels and lessons from marine hard substrata. *Australian Journal of Ecology* **23**, 8–26.

Drake J.A. (1991) Community-assembly mechanics and the structure of an experimental species ensemble. *American Naturalist* **137**, 1–26.

Harrison S. & Hastings A. (1996) Genetic and evolutionary consequences of metapopulation structure. *Trends in Ecology and Evolution* **11**, 180–183.

Hershey A.E., Pastor J., Peterson B.J. & Kling G.W. (1993) Stable isotopes resolve the drift paradox for *Baetis* mayflies in an Arctic river. *Ecology* **74**, 2315–2325.

Hughes J.M., Bunn S.E., Kingston D.M. & Hurwood D.A. (1995) Genetic differentiation and dispersal among populations of *Paratya australiensis* (Atyidae) in rainforest streams in southeast Queensland, Australia. *Journal of the North American Benthological Society* **14**, 158–173.

Lake P.S., Bayly I.A.E. & Morton D.W. (1989) The phenology of a temporary pond in western Victoria, Australia, with special reference to invertebrate succession. *Archiv für Hydrobiologie* **115**, 171–202.

Chapter 7 *Why do organisms occupy particular habitats?*

7.1 Finding a place to live

It is a warm summer's day on the South Island of New Zealand. In a small stream set in a mix of native bush and farmland, a giant kokopu (*Galaxias argenteus*) maintains a constant position close to the head of a small pool. The fish is about 15 cm long. In the sunlight, gold flecks down its flanks contrast clearly against a dark green background. Occasionally, it moves slightly to one side or to the surface to capture small insects drifting with the flowing water. Ahead of it faster water tumbles into the pool down a rocky cascade, and behind it the water slows so that the flow through the deep middle section of the pool is almost imperceptible.

This particular giant kokopu may have chosen to hold this position in the pool for several reasons. At the moment, it is the largest fish that is active within the pool, hence it can physically dominate all the other active fish. Periodically, it reaffirms its dominance by attacking smaller giant kokopu, forcing them to hold positions downstream in the slower section of the pool. The dominant fish is first in line for prey, the others have to wait their turn. Being the dominant fish it can select the position from which it can most efficiently catch its prey. It could move slightly upstream and into the faster water at the head of the pool where food would be delivered on the current more rapidly. However, maintaining a position in the faster water requires extra energy and prey is harder to capture as it drifts quickly past resulting in missed strikes.

It is now dusk. The fish we observed earlier is still feeding at the head of the pool. Behind, a larger giant kokopu slowly emerges from an undercut bank where it has spent the day resting. This fish is at least 25 cm long. Initially, it swims slowly up the pool, but then accelerates and charges at the 15 cm fish at the head of the pool. The larger fish chases the other fish twice around the pool. Finally in desperation, the smaller fish frantically swims into the fast cascade of water coming into the pool, then struggles up the cascade and into a small pocket of still water that lies several metres upstream. A second smaller fish that has also been chased from the pool by the large giant kokopu soon joins it. By day, these fish may choose their habitat on the basis of gaining efficient access to prey. However, by night their choice of habitat is forced on them by the need to avoid larger, more aggressive competitors.

7.2 What are habitats?

Different species live in different **habitats**, a word used to describe the local environment occupied by organisms. Such habitat descriptions usually include the obvious physical features of the local environment (e.g. pools and cascades in the opening example), and we have already introduced you to some habitats of freshwater systems. Recall from Section 4.4, for example, that we use the word littoral to describe the area of water and substrata close to the margins of lakes, whereas organisms living out in the open water above the bed are described as being in the pelagic zone. We may also use the word **micro-habitat**, which describes an organism's environment over distances of millimetres to centimetres to metres, scales smaller than those typically used for habitat. We include descriptions of micro-habitats because organisms may sometimes be affected by factors operating over small spatial scales as well as larger scales (see Chapter 2).

Within their habitats (and micro-habitats), organisms acquire the resources of food, living space, shelter and so forth that they require for growth, survival and eventual reproduction. Typically we will find some species have very broad diets and may occur in a variety of habitats – we call such organisms **generalists**. In contrast, we may find that other species live in only specific habitats (or even micro-habitats) and/or eat a very narrow range of foods, and we call these taxa **specialists**. The reasons why some organisms are specialists and others generalists lie partly within events long past – in the evolutionary history of each species (Box 7.1). Nevertheless, we can ask: What modern-day factors restrict some organisms to living in a small range of habitats or micro-habitats? Can this limitation explain why some species are common and others rare? These are central questions in population ecology, and we will now look at how ecologists have approached getting answers to them.

7.3 How do organisms end up in particular habitats?

Before we begin looking at different models that can explain species distributions, it is useful to break up the process by which organisms

Box 7.1

The role of history

One thing ecologists must always keep in mind is that many patterns we see in nature today may have been determined by events that happened long ago. For example, many species evolved on the super-continent Gondwana. Gondwana eventually broke up to form the major land masses of South America, Africa, Antarctica and Australia (as well as some smaller land areas). Plants and animals that descended from species that evolved on Gondwana often have distributions restricted to the above-named land masses – that is, we can explain some of the modern distribution of plants and animals on the basis of an event that happened a very long time ago. Climatic events have affected distributions markedly too. During the last ice age, parts of continental North America were covered by glaciers, which denuded the landscape of plants and animals except for isolated pockets. Today, we find

species that are restricted to mountain tops in some parts of North America, a distribution that is a reflection of this huge fluctuation in climate. In contrast, Australia was never affected much by glaciers, but its climate fluctuated markedly in other ways (such as becoming much drier). So, some of the differences we see between Australia and North America in modern-day distributions and diversity reflect their vastly different climatic histories quite independently of the effects of modern climatic regimes.

The reason why history is important is that the factors currently limiting the distributions of individual taxa (these are called **proximate factors**) could be quite different from the factors that affected their distributions in the past. An organism's current use of habitats may have evolved under environmental conditions very different from those operating now. So, when we ask questions about modern species distributions, we must think carefully about the historical context of those questions.

disperse and choose places to inhabit into several parts. This is because we can then see that factors influencing species distributions can operate at different points in this process. A young organism hatches, is born or dispersed from its parent – the first step can be dispersal away from the natal site (Fig. 7.1). As we discussed in the last chapter, some organisms may disperse long distances from their natal sites (perhaps

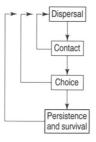

Fig. 7.1 Division of the process of habitat colonization into the separate steps of dispersal, initial contact, choice of specific habitat location, and post-choice persistence and survival. The arrows show the sequences in which events can occur. Various factors influence the success of individuals at finding and settling into habitat at each of these steps, and these factors are likely to vary among taxa and types of ecosystem. Individuals of some species (e.g. sessile taxa such as plants) may disperse and colonize new habitats only once during their lifetime, whereas other, highly mobile species may disperse and colonize habitats multiple times.

travelling between populations) and others very little at all. Following dispersal, organisms make contact with locations of potentially suitable habitat. They then may assess whether that habitat is suitable or not, using behaviour that is instinctive (i.e. genetically based) and choose whether to stay or to go. Plants cannot 'behave' in this sense of course, but seeds or spores may not germinate until they are in a suitable location or until a certain event has happened. Following selection of a suitable habitat spot, organisms must then persist and survive. It is possible that they may be driven away by competing species, eaten by predators or die because the habitat subsequently becomes unsuitable, so that after some time has passed the distribution of organisms does not reflect the choices they made during colonization. Some organisms may go through this sequence multiple times through their lives, dispersing and colonizing different habitat patches each time, whereas individuals of other species may get only one opportunity. Organisms that settle permanently in one spot, like plants and some animals (e.g. sponges), disperse only during juvenile stages, and habitat choices made by larvae (or germination by seeds and spores), if any, determine where adults can live.

Why is it useful to break up the process of how plants and animals find themselves in certain habitats into its component steps? Well, it allows us to see firstly that there is a wide array of possible factors influencing where an organism lives, and secondly that these factors can be viewed as a series of explanations of increasing complexity. Hence, a species may be limited to occupying particular habitats or micro-habitats simply because individuals are unable to disperse to some suitable locations. Alternatively, it may be limited to those places not because it has any difficulty in dispersing but because a superior competitor has already occupied some suitable places. The latter model is more complex because it requires knowledge about the effects of factors in all the steps preceding choice.

7.3.1 A simple model of habitat selection

One of the most straightforward models we can propose to explain species distributions is that of **simple habitat selection**. In this model, an individual is able to disperse to, and locate, all suitable locations in the area, and it prefers to inhabit particular places over others. No other species prevent it from occupying its preferred choices. If this model is correct, we should be able to predict where we will find this species in greatest abundance simply by knowing what kinds of habitat preferences individuals make. A simple habitat selection model thus focuses exclusively on the step where organisms choose and assumes that factors affecting dispersal, location or post-choice persistence and survival are all unimportant.

Box 7.2

Adaptation and Just So Stories

Charles Darwin's theory of natural selection is so well known that it is very common for people, from all walks of life, to describe some plant or animal feature (say of morphology or behaviour) as 'adaptive'. Usually, such stories describe why that feature causes the organism to 'be fit', and the implicit assumption is that all characteristics of an organism have been shaped by the mysterious 'forces' of natural selection. The difficulty is that such stories, when they are structured as arguments about why an organism possesses that feature or lives in a particular place, are inherently circular. That is, the story-teller assumes that natural selection has shaped all biological characteristics, thinks up a plausible way that this may have happened for that feature and the environment to hand, and often uses these explanations as a way of reinforcing the truth of the theory of natural selection. Of course, the theory is never in any danger of being discarded or disproved by this sort of reasoning! What seems on the surface to be a reasonable argument is nothing more than a rewording of the original premises (a fallacy that philosophers call a tautology or circular

argument). Such tales are sarcastically labelled by some evolutionary biologists as 'just so stories', after the Rudyard Kipling stories for small children, in which Kipling spins fantastic and amusing yarns about how leopards became spotted and camels got humps.

Does this mean we should never speak of 'adaptations'? No, but we should always be very careful *not* to place too much stock in inherently appealing stories. It is difficult to gather data that demonstrate that any particular feature is an 'adaptation'. Moreover, individual features that evolved in response to past environmental conditions long since changed (Box 7.1) almost certainly evolved in response to multiple environmental problems rather than just one, and indeed may have been shaped by events (such as founder effects and random genetic drift) other than natural selection *per se*. Another difficulty is that the theory of natural selection coupled with modern genetics is nowhere near as easy to understand, in all its details, as the popular press may lead you to believe. Terms like natural selection, fitness, adaptation and evolution all have specific meanings but are frequently misused, both by well-intentioned journalists as well as scientists without any training in evolutionary biology.

Researchers with simple survey data commonly invoke habitat selection to explain the distribution of organisms in nature. We should warn you, however, that some researchers fall into a trap of making circular arguments (Box 7.2). In these instances, researchers collect survey data to document the habitat use of species and conclude that 'preferences' determine where organisms are found. This sort of reasoning springs from an assumption that does not usually appear anywhere in their paper: that natural selection has produced behaviour that leads organisms to choose only places that provide optimal survival and reproductive success (i.e. behaviour is 'adaptive'). 'Adaptive preferences' are thus presumed to determine habitat selection and hence distribution. However, if all we have are survey data, the above 'conclusion' is nothing more than a restatement of the model (Box 7.2). That is, having carried out a survey and shown that some species exhibit specialized habitat use, we conclude that the places occupied by these organisms must be optimal because that is where they are found! This sort of reasoning is unfortunately common. The sheer number of these studies can give the impression that habitat selection is a well-supported explanation for distribution. Unfortunately, many of these papers do not provide evidence for or against any models of distribution, including that of simple habitat selection.

7.3.2 Other models that explain species distributions

A perusal of Fig. 7.1 illustrates the many other possible models we can consider besides that of simple habitat selection. The common feature of all these models is that some factor or factors prevent individuals from realizing all possible resources and habitats that are available. For example, organisms may be unable to disperse to some areas containing good habitat (see Chapter 6). Even if they are able to reach an area, they may not be able to find some suitable spots even when they are nearby. An organism's own behaviour can lead it to exclude some places that are suitable – possibly because that behaviour evolved under different environmental conditions (Box 7.1). Another suite of factors preventing organisms from occupying particular locations are interactions with other species – competitors for the same food for example, or predators that might be particularly effective hunters in some parts of the environment (see Chapter 9).

Judging from experiments carried out so far, the likeliest outcome is that distributions are determined by a combination of factors, not just one. The challenge now is to discover whether there are any generalities – that is, can we make predictions about the kinds of species or kinds of ecosystems where particular limiting factors will be important? This is an exciting question! We now turn to some examples that will show exactly how these different sorts of factors can work, using common stream organisms for illustration.

7.4 Blackfly larvae in streams

We shall look at a series of studies on some insects called blackflies. Blackflies are only a few millimetres long and larvae attach to hard substrata temporarily, which makes them amenable to experiments where we can manipulate and separate the effects of different factors affecting their distribution.

Blackflies (Order Diptera – the true flies) are common components of stream fauna around the world. Adult blackflies are small flies, and the females of many species bite and draw blood from mammals and other vertebrates. They are sometimes serious pests of both human beings and livestock and carry parasitic organisms that cause serious illnesses (e.g. river blindness caused by the parasitic nematode *Onchocerca volvulus*) in some parts of the world. Adult females lay their eggs in flowing waters, and the larvae make their way to **riffles** (places with fast-flowing,

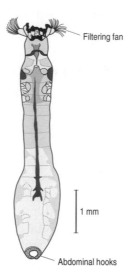

Filtering fan

1 mm

Abdominal hooks

Fig. 7.2 Drawing of a blackfly (Simulidae). The fans on the head are used to filter fine particles from the water column. The abdominal hooks are used to anchor the larvae to the substrate in fast-flowing water. (After Winterbourn *et al.* 2000 with permission of the Entomological Society of New Zealand.)

turbulent water flow). The larvae (Fig. 7.2) have a pair of fans which they use to filter particles as small as bacteria out of the water column. Periodically, each larva will flick each fan down and wipe particles off into its mouth. The larvae require hard substrata to attach to in order to filter feed, and to attach to the substratum they spin a small pad of silk that sticks to hard surfaces. The larvae have a circlet of hooks at the end of the abdomen, and they slip these hooks into the silk pad. This allows them to stand upright and face into the current. Not surprisingly, blackfly larvae require at least some water current to be able to filter out passing particles, and we might hypothesize that the faster the water, the more particles are likely to pass through their fans. This then is the micro-habitat sought out by blackfly larvae: hard substrata like rocks and logs, a relatively smooth surface for attachment and fast-flowing water that delivers small particles of food.

Larvae can use either of two dispersal methods. First, they can use the stream current to drift downstream. They sometimes trail a thread of silk out behind them while drifting, which snags onto hard objects and allows larvae to pull themselves down to the 'rock or log' surface. Second, larvae can move around the surface of individual substrata by looping, which is where they alternatively attach the mouth and abdomen to move over the surface (a bit like some caterpillars). Looping allows larvae to move small distances (millimetres to centimetres), which might allow them to move to nearby spots with faster current.

7.5   Blackflies and simple habitat selection

So, can we predict accurately the distribution of blackfly larvae in streams? Let us start with a model of simple habitat selection. Our knowledge of their habitat requirements would lead us to predict that if blackfly distribution is determined solely by habitat preferences, we will find blackflies in all areas of the streambed where there are hard substrata and the water is relatively fast flowing. Because faster water usually means more food per unit time, we would also predict that larvae will be more abundant in fast-flowing areas than in moderate or slower flows. (Note that it is possible to get slow water in riffles because of the zones of greatly reduced velocity created by flow around large objects like boulders.)

Are these predictions successful? The answer is, like many in ecology, yes – to a point. We usually find no (or very few) blackfly larvae in slow-moving areas of rivers like pools, backwaters or dead water zones, and often many larvae in riffles. That is consistent with our expectation that blackfly larvae choose places with water that is, overall, flowing reasonably fast. Nevertheless, some rivers and streams may have no blackfly larvae at all, even when the habitat appears very suitable. Part of the reason may lie simply in the evolutionary history of these species – perhaps blackflies never reached that particular stream (Box 7.1). Nevertheless, we should also remember a lesson we learned in Chapter 6. Species with complex life-histories can have stages living in quite disparate environments. The riparian habitat around rivers may be unsuitable for the requirements of adults, which means there will be no females to lay eggs and produce larvae. So we must always keep in mind that some of the explanation for the distribution of larvae may lie with the requirements of adults and, in particular, the egg-laying needs of females. (Likewise, if we are investigating the distribution of adults, we might find their occurrence largely set by the requirements of larvae!)

However, even if we found that blackfly larvae occurred in all riffles but not pools and dead water zones, that finding does not provide us with much insight. The result has confirmed a habitat type that is unsuitable – so it provides some explanatory power at that scale – but it does not necessarily tell us what sets limits to abundances in places that *are* suitable. To see why this is so, suppose we sample a number of riffles and find that abundances of blackflies vary greatly among them, or among fast-flowing spots within the same riffle. Can our model of simple habitat selection explain these differences? If so, it would mean that blackflies are able to, and do, choose between riffles (or specific stones) differing in average water velocity.

We will first provide an example of a study that is consistent with this explanation, then a series of far more comprehensive studies showing

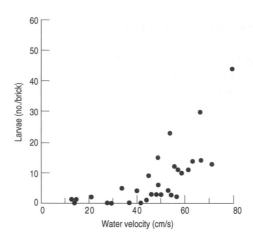

Fig. 7.3 The number of blackflies, *Austrosimulium victoriae*, occurring on substrata (bricks) having different water velocities. (After Downes & Lake 1991.)

that simple habitat selection is unlikely to be a useful model for explaining blackfly distribution. In our first example, Downes and Lake (1991) found that abundance of some Australian blackfly species varied greatly between individual stones within one riffle. They measured the average current speed over stones and then sampled them and counted the number of blackflies that were present. They found that numbers of blackflies were associated with the velocity of the water passing over those stones. Stones that were in relatively fast water had more blackflies than those that were in slow water (Fig. 7.3). So, the explanation for why blackflies are not found in pools (slow water velocities) helped explain within-riffle variability in this study. Possibly, a simple model of habitat selection (that blackflies can and do choose stones occupying relatively fast water velocities) explains the distribution of these species. We say 'possibly' because the consistency of observations with this hypothesis is not strong evidence for it. As explained above (Section 7.3.1), we need to collect further data that demonstrate that blackflies can and do have such preferences before we can conclude that simple habitat selection is a useful model.

We now turn to a series of studies that demonstrates very neatly how far more complicated nature is than this!

7.6   Blackfly larvae and other models of distribution

Hart *et al.* (1996) studied the distribution of larvae of the blackfly *Simulium vittatum* in Pennsylvania in the USA. Feeding rates of blackflies depend on the speed of water flowing through their fans. This knowledge led them to expect that, under a model of simple habitat selection, *S. vittatum* should be most abundant on stones within riffles that had fast flows. However, in contrast to the study of Lake and Downes (1991), numbers of blackfly larvae on stones were not associated well with water velocity over those stones (Hart *et al.* 1996). Numbers of blackflies were associated with flows when the latter were measured at much smaller spatial scales – at the scale of a millimetre, but this did not explain why stones in overall fast flows did not have more larvae than stones in slow flows.

Could this lack of association be a consequence of blackflies being prevented from reaching some locations during dispersal and that behaviour during dispersal changes the sorts of initial locations they are able to contact? In other words, can blackflies control their dispersal in the current and choose locations where to stop? These questions were addressed for *S. vittatum* by Fonseca (1999), who observed the behaviour of blackfly larvae in a specially constructed flume (flow tank) where relatively laminar flows could be generated. (Laminar flows have fairly predictable velocities at different heights above the bed and hence are more tractable to work with than turbulent flows, which can be chaotic and virtually unpredictable – see Section 5.5.) In the first experiment, Fonseca (1999) measured the rate at which both dead and living larvae sank in a simple glass tube, to see whether they can control their dispersal up and down through the water column. The sinking rates of dead and living larvae were the same, suggesting that blackfly larvae cannot control their positions in the water column, a behaviour that can greatly alter dispersal distances in streams. If larvae can maintain a high position in the water column, they can disperse further because velocities are generally higher at greater distances above the bed even in turbulent flows (refer back to Chapter 6). The fact that larvae sink without being able to do much about it means that they can be modelled essentially as dispersing passive particles. We should be able to use relatively simple mathematical equations that can make predictions about how far larvae will travel when dispersing from different heights above the bed.

This is what Fonseca (1999) did in the next experiments using the flume, where she released larvae at known heights in the water column and measured how far they travelled in different rates of flow before they were able to attach themselves to the substrate. The simple equations were largely successful at predicting these distances, except for the highest flow (Fig. 7.4). In high flows, larvae were unable to attach at all before being swept to the end of the flume, even when their silk line snagged on the flume's surface. These results suggested that larvae may be unable to attach to optimal sites in the field (remembering that fast

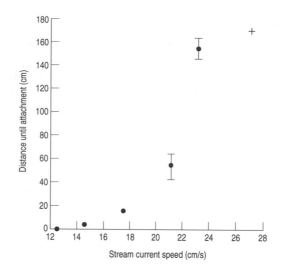

Fig. 7.4 Relationship between distance travelled in a flume by *S. vittatum* larvae until attachment (cm) and stream current speed. Distances travelled were well predicted by mathematical equations of the dispersal of passive particles in different water velocities. The distances travelled after releases at 27 cm/s (cross) were 170 cm and were not included in the analysis because no larvae were able to attach to the substrate. The error bars represent ± 1 standard error and there were 20 larvae used for each current speed. (After Fonseca 1999 with permission from Springer-Verlag.)

water flow equates with potentially greater food supply) simply because the water flow is too fast.

Fonseca and Hart (2001) confirmed this using a field experiment. They placed artificial substrata throughout a riffle of Taylor Run (Pennsylvania, USA). Larvae colonized these substrata and, consistent with previous observations, some substrata gained far more larvae than others, despite water flows over the substrata being similar (that is, numbers of larvae on substrata were not well explained just by water velocity). After noting which substrata obtained high numbers of larvae, Fonseca and Hart (2001) removed all blackflies from all substrata, and then reattached the *same* number of larvae to all of them. They then observed what happened to the numbers of larvae. Their reasoning was this: if substrata that originally had low numbers of larvae were low because they offered poorer habitat quality, then larvae attached to these substrata should leave at higher rates than larvae attached to substrata that previously had had higher numbers of blackflies. Thus, if the initial abundance patterns reflected habitat preferences, then the numbers of larvae should readjust to return them back to the high and low numbers originally recorded for those spots. Alternatively, if the numbers present on substrata were a reflection of

102

differences in initial settlement rates – driven by different exposures to numbers of settling larvae for example – and unrelated to habitat quality, then emigration rates should be comparable between these two sorts of substrata. Fonseca and Hart (2001) also monitored substrata that were cleared of all larvae at the start of the experiment, which allowed them to correct for new larvae settling onto their substrata during the experiment.

Fonseca and Hart's (2001) results showed that the emigration rates of larvae from substrata that previously had many larvae were equivalent to those that had few. This suggested that the reason that numbers differed was not because of habitat quality but because the rates of dispersal and settlement originally differed, as the flume experiments had suggested. Larvae of *S. vittatum* are an example of a species where constraints during dispersal and contact with the substrate restrict the number of places larvae are able to occupy. This is because their method of dispersal over long distances – using the drift – allows them little control over where they can land. They can only exhibit choice by leaving substrates – not by choosing the best ones in the first instance.

Do post-choice factors also affect the distribution of larval blackflies? Hart and Merz (1998) examined the effects that a predatory flatworm has on the distribution of larval *S. vittatum*. The flatworms live on the bottom of substrata or within tufts of algae, but venture out onto top surfaces to feed. In a small site on Chester Creek (Pennsylvania), blackfly larvae are their major prey. Do flatworms modify the post-attachment distribution of blackfly larvae through predation? In Chester Creek, blackfly larvae are most abundant in places where the water velocity is higher. However, flatworms are less abundant in high-velocity spots. To what degree is blackfly distribution determined by larval preferences for high water velocities compared with the effects of predation? Hart and Merz (1998) carried out an experiment using tiles, which were readily colonized by both blackflies and flatworms. They removed the flatworms from half of the tiles by dislodging them. Then they reduced the water velocity over half of these tiles and half of the undisturbed tiles by placing small mesh fences upstream of them. This experiment allowed them to examine the separate effects of predation and water flow upon blackflies. They found that blackflies had higher emigration rates from substrates where the water velocity was reduced, and that abundances were also reduced where flatworms were present (Fig. 7.5). The latter only ate about half the blackflies – the rest fled into the drift to escape. The experiment also suggested that flatworms were less effective as predators in high-velocity treatments, because they were rarely seen on the top surfaces of tiles in fast-flowing water. The authors concluded that the negative effect of water velocity on flatworms creates a refuge from predation for blackflies.

103

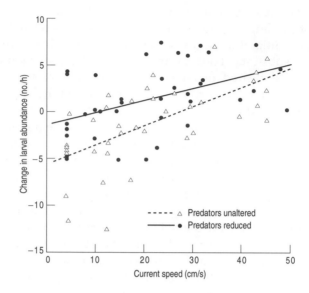

Fig. 7.5 Relationship between the rate of change in larval blackfly abundance and current speed in the reduced-predator (solid line) and unaltered predator (dashed line) treatments. The response variable was the change in larval abundance per hour. Statistical analysis showed that both predator abundance and current speed were significant predictors of changes in larval abundances. (After Hart & Merz 1998 with permission from Springer-Verlag.)

7.7 Conclusion

Organisms like *S. vittatum* demonstrate that high abundances in particular habitats can be caused by a variety of factors. *S. vittatum* larvae have limited ability to control the distance they disperse (Step 1, Fig. 7.1). Nor can they reliably contact optimal locations from the drift (Step 2, Fig. 7.1). This means that the highest quality habitats (i.e. stones with fast water flows) do not necessarily have the highest abundances of blackflies. Larvae can express habitat preference (Step 3, Fig. 7.1) by emigrating from poor locations, but stones with relatively high numbers of larvae may have only moderate flow. Additionally, the distribution of blackflies after settlement can be modified by the behaviour of predators (Step 4, Fig. 7.1). The flatworms were more likely to disturb and capture blackflies in slower velocities. In that case, the behaviour of the predator may be at least as important as the current speed in determining distribution of the blackflies.

The series of experiments described above demonstrates how complex things can be. A relatively naïve person observing that blackflies are more abundant in higher flows might conclude that the simple habitat selection model obviously applies. That conclusion could very well be a consequence of a belief that distributions are inevitably determined by adaptive habitat preferences (Box 7.2). Hopefully now, you can see the pitfalls of such a simplistic approach and how an elegant series of experiments can be used to tease apart multiple hypotheses (see Chapter 1).

7.8 Further reading

Downes B.J & Keough M.J. (1998) Scaling of colonization processes in streams: parallels and lessons from marine hard substrata. *Australian Journal of Ecology* **23**, 8–26.

Downes B.J. & Lake P.S. (1991) Different colonization patterns of two closely related stream insects (*Austrosimulium* spp.) following disturbance. *Freshwater Biology* **26**, 295–306.

Fonseca D.M. (1999) Fluid-mediated dispersal in streams: models of settlement from the drift. *Oecologia* **121**, 212–223.

Fonseca D.M. & Hart D.D. (2001) Colonization history masks habitat preferences in local distributions of stream insects. *Ecology* **82**, 2897–2910.

Hart D.D., Clark B.D. & Jasentuliyana A. (1996) Fine-scale field measurement of benthic flow environments inhabited by stream invertebrates. *Limnology and Oceanography* **41**, 297–308.

Hart D.D. & Merz R.A. (1998) Predator–prey interactions in a benthic stream community: a field test of flow-mediated refuges. *Oecologia* **114**, 263–273.

Kipling R. (1950) *Just So Stories: For Little Children*. Library Edition, Macmillan & Co., London.

Winterbourn M.J., Gregson K.L.D. & Dolphin C.H. (2000) Guide to the aquatic insects of New Zealand. *Bulletin of the Entomological Society of New Zealand* **13**, 102p.

Chapter 8 *How do freshwater communities recover from disturbance?*

8.1 A storm in Fiordland, New Zealand

On a fine sunny day, the view from the shores of Milford Sound is simply stunning. Dense *Nothofagus* beech forest covers the lower slopes of the surrounding mountains. Barren rocky cliffs and white snowfields rise spectacularly above the forest (Fig. 8.1). However, even in this tranquil scene, the damage wrought by the severe rainstorms that are generated as the moisture-laden winds of the 'roaring forties' slam into the Fiordland mountains is clearly marked on the surrounding forests and along the small streams that flow into the Sound. Occasional barren bedrock scars run down the mountain sides and on for hundreds of metres into the forest, the result of huge land slips that hurtled tonnes of mud, rock and water down from the higher slopes. Along the streams lie huge boulders and the trunks of smashed trees that were carried down the slopes during past storms.

Fig. 8.1 Milford Sound in south-west New Zealand on a calm day. Patches of exposed bedrock reveal the scars of past land slips.

How do freshwater communities recover from disturbance?

There has not been a major storm at Milford Sound for over 5 years, and a small stream runs as a gentle trickle amongst moss encrusted boulders, small trees and shrubs down into the Sound. In the stream itself, *Deleatidium* mayflies run over the rocks, grazing on the diatoms that grow in the bright sunlight. The bed of the stream is a stable habitat that provides abundant food and shelter from the predatory fish such as koaro (*Galaxias brevipinnis*) and banded kokopu (*G. fasciatus*) that swim above. Many minor floods have washed down the stream in the past 5 years, but nothing has moved the largest rocks since the last major storm. Today, wispy cirrus clouds begin to streak across the sky, warning of the approach of yet another cold front and associated rain. However, rain falls every few days in Fiordland and the approach of another cold front generally passes without comment.

On the first day over 140 mm of rain falls, a total that is not unusual for Fiordland over a 24-h period. However, unlike most storms, the rain keeps falling as the cold front stalls over Fiordland. Next day over 170 mm pours down, with a massive 320 mm falling on the following day. Torrents of water run down the mountains and into the streams. As water levels rise, so does the velocity of the water. The mayflies sense the increasing water velocity and move to shelter under the rocks. The water continues to rise and the velocity of the water increases further. The mayflies move deeper into the substrate, perhaps warned of the increasing force of the flood by the vibrations generated by the rushing water, or perhaps by the rattle of rocks beginning to slip along the bed of the stream. At first, only an occasional rock moves. However, the movement of one rock loosens surrounding rocks. They too slip, and in doing so, trigger more rocks into motion. The stream soon turns into a cascade of rolling and grinding rocks and logs. The rocks under which the mayflies took shelter are now also moving, and the mayflies are themselves tumbling downstream in the rushing water. Those mayflies not crushed between the rolling rocks are soon washed out of the forest stream and to their death in the deep waters of Milford Sound.

After 4 days, the rain stops and the sun shines once again. New bedrock scars run down the mountains, and trees ripped from the mountains by the force of the falling water float in Milford Sound. Flows rapidly drop back to a trickle, however the stream now runs down through a wide swathe of destruction. Where bank side trees and shrubs previously grew, only beds of sterile sand and freshly dumped boulders lie. In the stream, only the water moves. Rock surfaces are clean, having been scoured of any algae as they rolled along the stream. Not a single fish or mayfly can be found, and it looks as if the devastated stream community will never recover. However, such storms have happened before. Recovery will take time, but it will happen.

8.2 How do communities recover from disturbance?

Our observations of disturbances such as floods lead to predictions of how we might expect a community to respond to a specific disturbance (Box 8.1). We can predict that a disturbance will result in reductions in the abundance and diversity of organisms, as individuals are removed from the habitat due to the unfavourable conditions. When favourable conditions return, a process of community rebuilding will commence, as the remaining individuals reproduce or new individuals migrate into the habitat from elsewhere (see also Chapter 6). The numbers, biomass and diversity of the community will initially increase from the low levels that were present after the disturbance. Rates of increase will eventually slow or even level off as the community completes its recovery, although the period of time required to rebuild the community will depend on the

---

Box 8.1

What is disturbance?

Loosely, a 'disturbance' refers to some form of event that disrupts the normal functioning of a community. Severe floods are only one of many types of disturbance that may occur in freshwater systems. However, a disturbance may come in many forms, including drought, periods of extreme heat or cold, or spills of pollutants. Each type of disturbance also varies with respect to its spatial and temporal extent, intensity and timing. A pond may only partially dry, leaving a central pool of water that can act as a refuge for species that cannot tolerate drying. Alternatively, a major drought may result in the drying of all ponds across an entire region, significantly depleting the sources of potential colonists that might be available when the ponds refill. In streams, substrate disturbance may occur on a small or massive scale. A single rock may roll, dislodging all animals and providing local opportunities for recolonization. In contrast, a major flood, such as that described at the start of the chapter, can scour many kilometres of stream and completely eliminate the plant and animal communities that normally reside there.

Disturbance removes individuals from a community. Which individuals are removed will vary depending on the intensity of the disturbance, the capacity of any one individual to tolerate the harsh conditions that prevail during a disturbance event, and luck. Even the strongest individuals may be in the wrong place at the wrong time when a rock rolls over. Common species within a community may be the most severely impacted during a disturbance, simply because there are more of them to be affected. As a result, disturbance can play a role in preventing habitats from being taken over by a limited number of highly competitive, dominant species. Disturbance may therefore play a role maintaining species diversity. Alternatively, frequent severe disturbance can result in very low species diversity with few species being able to survive the repeated impact of the harsh conditions.

The recovery of a community from a disturbance is a process known as succession. The nature of any particular succession will be influenced by many variables, including the spatial and temporal extent and intensity of the disturbance, the physical and chemical conditions that prevail after the disturbance, the composition of the individuals that survived the event, and the ability of potential colonists to locate, colonize and reproduce in the previously disturbed habitat. Again, a random element is introduced into the processes regulating community structure. Just as an individual may be removed from a community by being in the wrong place at the wrong time, another individual may fortuitously settle in a recently disturbed area where the numbers of competitors are reduced and resources are temporarily abundant.

intensity and spatial and temporal scales across (see Chapter 2) which the disturbance occurred. Community composition may alter through time as new colonists arrive, and various competitive and predator–prey interactions develop. Close observation of natural events can produce numerous useful predictions. However, such predictions can only be tested through the detailed study of real disturbances such as droughts and floods.

8.3 Flash floods in desert streams

A major difficulty in studying the impact and recovery from disturbance is that disturbance by its very nature tends to be unpredictable. Disturbances are events that are outside the range of what we might consider 'predictable and normal'. Consequently, organisms that have evolved to cope with 'predictable or normal' conditions are removed from the habitat when 'unpredictable or abnormal' events occur. Such events are logistically difficult to study as the researchers and their equipment have to be in the right place at the right time. Frequent natural disturbance does occur in the intermittent streams, such as Sycamore Creek, that drain the Sonoran Desert in Arizona. Sycamore Creek has been the focus of two decades of intensive research on the role of disturbance in lotic systems. In summer, Sycamore Creek often dries up apart from sections where bedrock lies beneath the sediments that form the streambed, forcing groundwater to the surface. Water trickles along the surface of the stream channel for a short distance downstream of these areas, before seeping back into the desert sand. In the absence of any rainfall, flow is low and constant along these wetted sections. When occasional thunderstorms occur, floodwater may surge down the otherwise dry channels. The rate of stream discharge at one of the normally stable permanent sections may rise 100-fold or up to 1000-fold over a period of just a few minutes as the flood wave passes, then recedes to pre-flood levels within a day. For a community accustomed to a gentle trickle of water, such floods have the potential to be catastrophic. However, in the warm and highly productive environment of a desert stream, the potential for rapid recovery is high. The combination of a frequent natural disturbance, combined with the potential for rapid recovery creates the ideal situation for the study of disturbance and succession.

8.3.1 Describing patterns of succession in Sycamore Creek

In early August 1979, a series of thunderstorms rumbled over the Sycamore Creek catchment, and sent several flash floods tumbling down the stream that, at most, had previously been a gentle trickle (Fig. 8.2). Within a day of the last of three floods, flow levels were back to near normal, and a team of freshwater ecologists began a pioneering study of community responses and recovery from the effects of the flooding (Fisher *et al.* 1982). Post-flood changes in the physical, chemical and biotic features of the stream were measured for 63 days after the flood. The study was finally terminated when a second flash flood occurred, ending the period of community recovery.

The flood event studied by Fisher *et al.* (1982) receded quickly, and by the time measurements commenced, many of the parameters had settled back to typical inter-flood states. Most of the physical and chemical parameters measured following the flood exhibited only small changes over the post-flood study period (Table 8.1). Parameters associated with stream flow (width, depth, surface area, velocity and discharge) declined as the flood-waters receded. Most of the chemical variables measured also showed little change over the post-flood period with the notable exceptions of nitrogen and phosphorus. Both nutrients were elevated immediately after the flood, then declined markedly over the next 13 post-flood days to low stable levels.

Biota recovered rapidly from this disturbance in the relatively warm (22–30°C) waters of Sycamore Creek. Diatoms colonized the disturbed habitat within 2 days of the flood, and covered close to 100% of the habitat within 13 days (Fig. 8.3a). Their abundance then declined markedly as blue–green algae and the filamentous algae, *Cladophora glomerata*, began to colonize the habitat 21 days after the flood (Fig. 8.3a). Fisher *et al.* (1982) estimated that less than 4% of the original invertebrates remained after the flood. Following the flood, invertebrate abundance increased albeit at a slower rate compared with the algal community. Within 13 days, biomass had attained 50% of the final biomass observed at the end of the study (Fig. 8.3b). Invertebrate numbers reached 50% of the final number

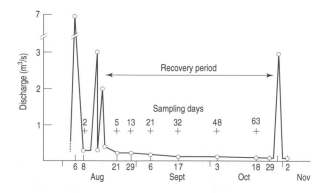

Fig. 8.2 Stream hydrograph and sampling dates on Sycamore Creek from August to November 1979. (After Fisher *et al.* 1982 with permission of the Ecological Society of America.)

110

Table 8.1 Morphometric, physical and chemical features of Sycamore Creek during post-flood recovery. Concentrations in milligrams per litre. (After Fisher *et al.* 1982 with permission of the Ecological Society of America.)

Date	8 Aug	21 Aug	29 Aug	6 Sept	17 Sept	3 Oct	18 Oct
Days after flood	**2**	**5**	**13**	**21**	**32**	**48**	**63**
Width (m)	7.2	6.6	6.3	6.0	5.7	5.4	5.2
Depth (cm)	6.3	7.2	6.4	6.4	5.7	4.4	4.9
Surface area (m^2)	3600	3300	3120	3000	2850	2680	2600
Discharge (m^3/s)	0.28	0.22	0.2	0.18	0.1	0.1	0.11
Mean velocity (m/s)	0.62	0.46	0.5	0.47	0.31	0.42	0.41
Temperature	29.6	26.3	27.3	30.2	28.5	24.4	22.4
Conductance (μS/cm)	610	630	560	540	530	570	560
pH	7.7	8.6	8.0	7.9	8.0	7.9	8.0
NH$_3$ N	0.017	0.005	–	–	0.014	–	–
NO$_3$ N	0.44	0.140	0.120	0.075	0.070	0.100	0.087
Soluble reactive phosphorus	0.094	0.065	0.070	0.062	0.056	0.054	0.056

after 35 days. The flood had only a relatively small impact on overall diversity of macroinvertebrates. A total of 48 macroinvertebrate taxa was collected over the study, with 38–43 taxa being present on any particular sampling date. The flood initially resulted in the net loss of five taxa, all of which subsequently reappeared.

Detritus derived from terrestrial (allochthonous) sources was the primary food of the invertebrate community immediately after the flood (Fig. 8.3c). However, within 2 weeks, diatoms dominated invertebrate diets. Later, detritus again began to dominate invertebrate diets, however, this time much of it was derived from algal (autochthonous) sources rather than terrestrial sources (Fig. 8.3c).

8.3.2 Are all floods the same?

Responses of the desert stream community to the flood studied by Fisher *et al.* (1982) largely corresponds with our predictions as to how communities may respond to disturbances such as floods. The physical and chemical environment was greatly altered during and immediately after the flood. Stable low-flow conditions resumed within a few days of the last flood. As predicted, the abundance and biomass of organisms (both algae and macroinvertebrates) were low immediately after the flood, but then increased rapidly as the stream discharge dropped and stabilized. Dramatic changes through time in the taxa composition of biofilm algal communities also occurred. However, not all of our predictions were accurate. The flood had only a limited impact on the number of macroinvertebrate taxa present in the stream, and macroinvertebrate abundance and biomass was continuing to increase 63 days after the flood.

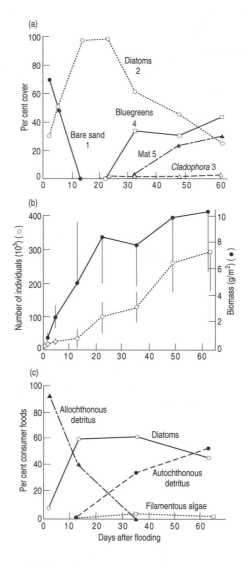

Fig. 8.3 Per cent cover of algal patch types (a), invertebrate biomass and numbers (b), and proportions of foods in invertebrates' guts (c) after flooding in Sycamore Creek. (After Fisher *et al.* 1982 with permission of the Ecological Society of America.)

The study by Fisher *et al.* (1982) does have some limitations, the most obvious being that it is based on a single flood in a single stream. Would a more severe flood have had a greater impact on taxa diversity? Would the invertebrate abundance and biomass have continued to increase had the recovery period been longer? The absence of samples before and after floods precludes the possibility of comparing pre- and post-flood communities. Consequently, we do not know if the community before the flood was similar to or different from the community that developed after the flood.

8.3.3 Are patterns of recovery following floods predictable?

Limited resources frequently force us to decide between collecting detailed information from a single event, or collecting less detailed information from a larger number of events. It is rarely possible to do both, and in the case of Fisher *et al.* (1982), the former approach was chosen. They measured a wide range of potentially important community responses, thus enabling identification of the variables likely to provide the most useful information. Future studies can then focus on a smaller subset of key response variables, thus enabling an expansion of the spatial and temporal scope (Chapter 2) of those subsequent studies.

Grimm and Fisher (1989) addressed the limitations inherent in a study based upon a single flood by studying multiple flood and post-flood periods in Sycamore Creek over 3 years in winter, spring and summer. In this subsequent study, they examined 20 post-flood recovery sequences (at 1-week intervals) of invertebrates and periphyton. The floods studied ranged from 0.2 to 58 m^3/s compared with the single flood of 7 m^3/s studied by Fisher *et al.* (1982). A range of recovery periods up to 158 days was studied, compared with the single 63-day period studied by Fisher *et al.* (1982).

The study of multiple flood events allowed Grimm and Fisher (1989) to examine the capacity of different elements of the stream community to resist floods of varying magnitude. They observed that desert stream communities are invariably sensitive to flood disturbance, with even small floods (0.2 m^3/s) resulting in community disruption (Table 8.2). Not all elements of the community responded in the same way. Cyanobacteria or blue–green algae densities were severely reduced by any flood, whilst floods of 8 m^3/s or greater were required to reduce densities of diatoms and invertebrates to less than 90% of pre-flood levels. In many cases, immediate post-flood densities of diatoms were higher compared with pre-flood densities, suggesting enhanced diatom production immediately after a flood.

The pattern of community succession was similar in winter or summer, with only some relatively minor differences observed up to 70 days post-flood (Fig. 8.4). Diatoms began to colonize substrates within hours of a flood event, followed by green algae and cyanobacteria. Increases in algal abundance was reflected by increasing chlorophyll a and ash-free dry mass (AFDM), measures of the quantity of algae and organic matter present in biofilm, respectively (Fig. 8.4a,b,e,f). Recovery of biofilm densities was slower in winter compared with spring–summer. Invertebrate densities also exhibited rapid recovery with chironomid midge larvae initially colonizing denuded substrates followed by oligochaete worms,

Fundamental Ecological Questions

Table 8.2 Resistance to floods (as per cent change across an event) of major periphyton groups and invertebrates in Sycamore Creek, Arizona, 1984–87. Floods are listed in order of magnitude (Q_{max} = maximum discharge). For periphyton, resistance reflects change in chlorophyll a of each type, while for invertebrates resistance reflects change in density. (After Grimm & Fisher 1989 with permission of JNABS.)

Q_{max}	Bacillariophyceae	Chlorophyta	Cyanobacteria	Invertebrates
0.2	+63	+82	−100	−
0.3	+147	−70	−100	−65
0.3	+[a]	−43	−68	−
0.5	+22	−34	−55	−
1.0	+	−87	−75	+137
1.1	−31	−77	−100	−
1.2	−75	+344	−100	−
1.7	+3811	−64	−100	−3
1.9	+1035	−95	−99	−
2.1	+146	−100	−100	−
2.3	+	−100	−100	−
2.4	+388	−73	−94	−64
2.4	−66	NP[b]	−100	−91
3.3	+	−100	−82	+139
3.5	−66	−95	−100	−98
4.9	−80	NP	NP	−
6.3	+295	−100	−100	−83
7.2	+431	−94	−100	−99
8.5	−99	−100	NP	−97
9.9	−100	−100	−100	−97
19.0	−27	−100	NP	−
19.6	+41	NP	NP	−
23.0	−79	−100	−100	−
23.4	−91	−100	−100	−21
24.7	−68	−100	−100	−
46.9	−95	−100	−100	−
58.4	−86	−100	−100	−92

[a]Post-flood appearance of algae which was absent before the flood. [b]NP, not present before or after flood.

then mayflies (Fig. 8.4c,g). Initial rates of recovery were similar in winter or summer, although the time taken for total macroinvertebrate densities to exceed 40 000 individuals per square metre was significantly faster in summer compared with winter. The maximum number of individuals and biomass ultimately attained in summer was lower in summer compared with winter (Fig. 8.4c,d,g,h).

The general patterns of community recovery described by Grimm and Fisher (1989) are similar to the patterns of succession observed by Fisher *et al.* (1982), at least up to 60–70 days post-flood. However, Grimm and Fisher (1989) were able to study several post-flood sequences of between 70 and 158 days in summer, significantly longer than the single post-flood period of 63 days studied by Fisher *et al.* (1989). Post-flood sequences up to approximately 70 days usually showed a trend of either increasing or stable invertebrate densities and biomass, a pattern that corresponds with the pattern observed in the flood studied by Fisher *et al.* (Fig. 8.4). However, for longer post-flood periods Grimm and Fisher (1989) observed a dramatic decline in invertebrate abundance and biomass (Fig. 8.4g,h). Crashes in macroinvertebrates were followed by

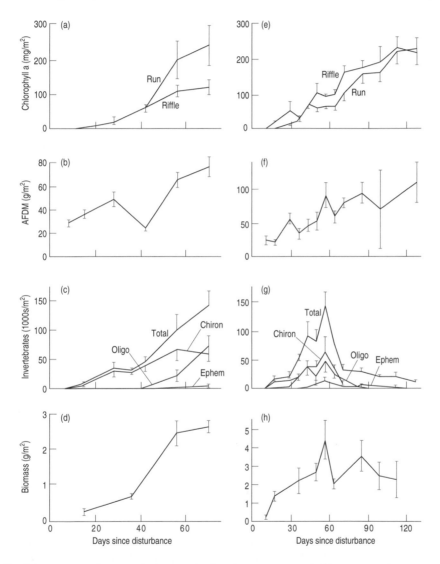

Fig. 8.4 Successional patterns for a winter (a–d) and spring-summer (e–h) successional sequence of chlorophyll a (a,e), ash-free dry mass (AFDM) (b,f), macroinvertebrate density (c,g), and macroinvertebrate biomass (d,h) following spates on Sycamore Creek, Arizona. Means ± standard error, $n = 5$. Chiron, chironomid midge larvae; Ephem, ephemeropteran mayfly nymphs; Oligo, oligochaete worms. (After Grimm & Fisher 1989 with permission of JNABS.)

increases in the abundance of nitrogen-fixing cyanobacteria or blue–green algae suggesting that nitrogen may have been limiting in the period following the macroinvertebrate population crash.

8.4 Models of community recovery in desert streams

Detailed descriptive studies, such as those conducted by Fisher *et al.* (1982) and Grimm and Fisher (1989), can be used to describe ecological events that are rich in patterns and apparent relationships between different variables. Such studies are a crucial step in developing models that describe the functioning of ecological systems. Based on the evidence gained from the study of a relatively large number of floods, Grimm and Fisher (1989) were able to propose a model of community recovery following flood disturbance in desert streams. They argued that desert stream communities are characterized by a pattern of low **resistance** and high **resilience** to flood disturbance. That is, even comparatively small floods will severely disrupt desert stream communities (low resistance), but recovery will be relatively rapid (high resilience).

Factors they believe contributed to the low resistance of the community to disturbance included the severity of floods, substrate instability and the scouring action of transported sediment. The high community resilience, as measured by the rapid recovery, was attributed to the warm temperatures, stable low-flow periods following floods and the high light environment that allowed biofilms dominated by diatoms to develop and sustain high levels of primary production. The high levels of nitrogen present in the stream enhanced diatom production in the immediate post-flood period. Subsequent dominance by blue–green algae occurred when nitrogen became limiting, and the nitrogen-fixing blue–green algae gained a competitive advantage. Grimm and Fisher (1989) also argued that nitrogen limitation was also likely to be the cause of the sudden declines in invertebrate abundance that occurred over long post-flood sequences (for a more complete outline of their model, see Grimm 1993).

8.5 Testing the flood disturbance model

Grimm and Fisher's model of flood disturbance and community recovery in a desert stream is inferred from their detailed observation of many floods in Sycamore Creek. Their model makes clear predictions as to how the various components of a desert stream community will respond to a flood, and they suggest likely causal relationships between certain processes and patterns. However, as discussed in Chapter 1, the inference of strong causal links between process and pattern requires caution, particularly when our inferences are based on correlations observed in descriptive data. Their model is however extremely valuable, as it offers numerous opportunities to generate hypotheses that can be tested experimentally.

The rapid growth of biofilms dominated by diatoms is a key feature of Grimm and Fisher's model of ecosystem recovery following floods in desert streams. Hence, understanding the mechanisms that influence biofilm growth is central to understanding the recovery process. During floods, the movement of rocks along the streambed scours biofilm from the substrate, leaving bare surfaces that are recolonized when stable post-flood conditions become established. Creating 'artificial floods' by rolling rocks around, or by placing bare rocks into the stream would seem to be a valuable tool for experimental studies of communities' recovery from floods. Individual rocks form discrete experimental units that can be readily replicated. Numerous factors can potentially be manipulated, such as the intensity of the disturbance or perhaps the nutrient levels around a particular rock. We could do this by varying the number of times we rolled a rock with established biofilm, or perhaps adding nutrients to the water that flows past a previously disturbed rock.

Realism is critical in any worthwhile experiment, and simply rolling a rock in a stream is not, unfortunately, a real flood. If we roll a rock in a stream that is not subject to flooding, then only the rock moves. The disturbance we have created is very local in its effect, being restricted to the biofilm and invertebrates on the rock that we are studying. However, as discussed in Chapter 2, processes occurring at a small scale may be greatly influenced by factors operating across much larger scales. In a real flood, processes operating at a local scale are also being influenced by environmental changes that may be occurring across the entire catchment. Before we can confidently use small-scale disturbances to study the processes occurring in a real flood, we need to know whether the mismatch in the scale of the disturbance that occurs between real large-scale floods and small-scale disturbance experiments really matters.

8.5.1 Does the spatial and temporal extent of a disturbance influence community recovery?

Do the environmental changes that occur across an entire catchment as a result of a flood influence patterns of community recovery? Would the algae communities develop in the same way if rocks were placed in Sycamore Creek either just after a flood, or perhaps 30, 50 or even 80 days after a flood? Peterson (1996) compared patterns of algal colonization and algal community development in biofilms on small ceramic tiles (28 cm^2) placed in Sycamore Creek in late June, 87 days after an April flood. Sampling was repeated 4 days after an August flood. Measurements of instantaneous rates of algal colonization were determined by incubating tiles in the stream for 3 h for 12 consecutive 3-h periods (36 h). The cumulative development of biofilm over the same 36-h period was measured by placing tiles in the stream at the start of the 36-h period, and then analysing the biofilm composition on three replicate tiles which were collected every 3 h over a 36-h period. If local processes determine the development of biofilm, then the development of algal communities on rocks will show no association with changes in the wider environment. Alternatively, if larger scale processes influence biofilm development, the development of algal communities in biofilms will vary with respect to the time since the last flood, and show some association with wider changes occurring in the environment as the entire community recovers from the flood-related disturbance.

Peterson (1996) measured the potential influence of the wider environment on the biofilm development in two ways. The abundance of potential algal colonists, represented by the abundance of algae drifting in the water column, was assessed by collecting hourly water samples (also see Chapter 6 for other studies examining the role of dispersal and colonization in community dynamics). Differences in the wider chemical environment in which the rocks were immersed was assessed by hourly measurement of the concentrations of nitrate nitrogen, soluble reactive phosphorus and conductivity. Both the abundance of drifting algae and water chemistry are influenced by processes operating upstream of the experimental tiles.

Peterson's observations of water chemistry and algal drift rates suggested that the wider environment was very different during the June and August study periods. Consistent with previous studies of Fisher *et al.* (1982), and Grimm and Fisher (1989), concentrations of nutrients (nitrogen and phosphorus) were significantly higher immediately after the flood (August), compared with the samples collected after a long inter-flood period (June) (Fig. 8.5). The density and composition of drifting algae in June compared with August were also markedly different (Figs 8.6 & 8.7). In June, overall algal drift densities were lower compared with August. In June, blue–green algae dominated drifting algae, whereas in August, diatoms dominated. Consequently, the tiles placed in the stream just after the flood were exposed to high numbers of potential colonists dominated by diatoms, and a high nutrient environment.

Patterns of algal colonization were also significantly different in June compared with August (Figs 8.6 & 8.7). In June, rates of colonization on the ceramic tiles were relatively low compared with rates of colonization in August. After 36 h of colonization in June, the community was dominated by blue–green algae. The actual cumulative rate of colonization was, however, less than what was predicted from analysis of the tiles placed in the stream for 3 h. This suggests that many of the cells that settled

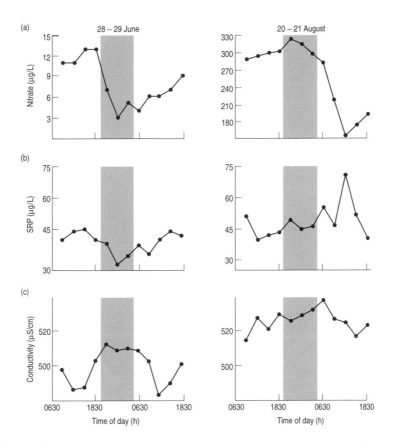

Fig. 8.5 Diel changes in (a) nitrate nitrogen, (b) soluble reactive phosphorus (SRP) and (c) conductivity in Sycamore Creek during June and August algae colonization studies. Shaded regions indicate night. (After Peterson 1996 with permission of Oikos.)

subsequently drifted away or died. Diatoms were present in the drift. However, in the low nutrient environment present in June, they did not appear to be capable of reproducing rapidly. In contrast, algal colonization in August was rapid, producing a biofilm dominated by diatoms after 36 h. Significantly, the cumulative rate of colonization after 36 h in August was well in excess of that which might have been predicted from rates of instantaneous settlement. Algal cells that settled subsequently began reproducing rapidly.

8.6 Does disturbance influence freshwater communities?

Twenty years of study on Sycamore Creek indicate that flood disturbance is a key factor regulating community structure of desert streams.

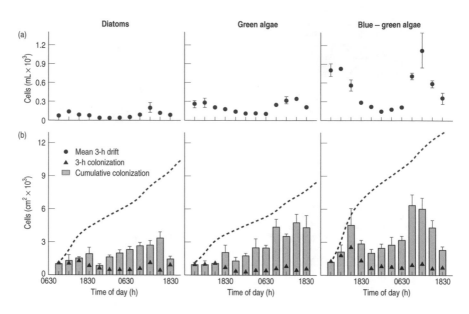

Fig. 8.6 Diel changes from 28 to 29 June (pre-spate) in mean algal cell density (± standard error) in the drift (a) (closed circle), and (b) on 3-h instantaneous colonization tiles (lower panels: closed triangles), and cumulative colonization tiles (lower panels: open bars) for each of the three main algal divisions. Dashed line plots the expected cumulative cell densities, calculated as the cumulative sum of sequential 3-h instantaneous colonization densities. (After Peterson 1996 with permission of Oikos.)

Flooding involves a change in conditions across an entire catchment, and these large-scale processes have a strong influence on community dynamics at smaller scales. However, whilst disturbing a small patch is not comparable to a real flood, comparing algal colonization on tiles during post-flood and inter-flood periods does provide valuable information on the processes involved in redevelopment of a community after a real flood. In Sycamore Creek, the high levels of nutrients present after floods do appear to play a key role in the rapid recovery of biofilm and primary production after floods. Diatoms are present in the drift at varying densities at all times, yet only reproduce rapidly for a few days after a flood. This hypothesis has been partially tested by Grimm and Fisher (1986) who found that the addition of nutrients does enhance rates of algal growth when nitrogen levels are low.

The studies of the processes influenced by disturbance in Sycamore Creek have contributed to the development of general models of the role of disturbance and community succession in lotic systems. However, work in Sycamore Creek also indicates that we should be cautious when extrapolating the results of one study across wider spatial and temporal scales. Whilst there are common processes involved in all

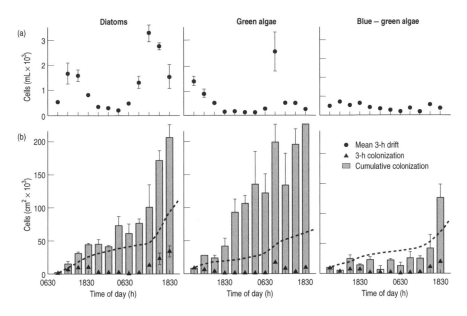

Fig. 8.7 Diel changes from 20 to 21 August (post-spate) in mean algal cell density (± standard error) in the drift (a) (upper panels: closed circles), and (b) on 3-h instantaneous colonization tiles (lower panels: closed triangles), and cumulative colonization tiles (lower panels: open bars) for each of the three main algal divisions. Dashed line plots the expected cumulative cell densities, calculated as the cumulative sum of sequential 3-h instantaneous colonization densities. (After Peterson 1996.)

floods, we can see from the variation in Sycamore Creek that factors such as flood intensity and the length of inter-flood periods can strongly influence community structure in unexpected ways. Further studies in Sycamore Creek have revealed that seasonal variation in conditions also influences succession following floods (Boulton *et al.* 1992) and that regulation of community dynamics due to nutrient limitation can show great spatial variation within the streambed (Fisher *et al.* 1998, Dent & Grimm 1999).

Extrapolating from models of flood disturbance in Sycamore Creek to other lotic and lentic systems, or perhaps to predicting the impact of other types of disturbance must be done with considerable caution. Whilst there are likely to be patterns related to disturbance and recovery from disturbance common to all systems, important differences are also likely to exist. Compared with Sycamore Creek, the impact of flood disturbance is likely to be quite different in streams where the substrate is relatively stable. Large cobbles, boulders or bedrock may not move in floods and could act as refuges from the disturbance (Matthaei *et al.* 1999). Such islands of stability may dramatically change the dynamics of recolonization and recovery from disturbance, and contribute to the spatially and temporally complex and often patchy distribution of organisms in streams (Townsend 1989). Rates of recovery may also be very different in a heavily shaded, low temperature stream running through a cool temperate rainforest.

The role of physical disturbance in structuring lentic communities has received considerably less attention compared with lotic systems. At present we know relatively little of the responses of lentic communities to the various forms of disturbance they may encounter. Wave action (McCall & Soster 1990, Rasmussen & Rowan 1997), drought (Jeffries 1994), seasonal drying (Lake *et al.* 1989) and bioturbation (Ten Winkel and Davids 1985, Breukelaar *et al.* 1994) have been shown to have significant impacts on lentic community dynamics. However, with respect to most of these forms of disturbance, our knowledge is restricted to a single system or disturbance event, and we have little idea as to how response patterns may vary at different spatial and temporal scales.

8.7 Further reading

Boulton A.J., Peterson C.G., Grimm N.B. & Fisher S.G. (1992) Stability of an aquatic macroinvertebrate community in a multiyear hydrologic disturbance regime. *Ecology* **73**, 2192–2207.

Breukelaar A.W., Lammens E.H.R.R., Klein Breteler J.G.P. & Tatrai I. (1994) Effects of benthivorous bream (*Abramis brama*) and carp (*Cyprinus carpio*) on sediment resuspension and concentrations of nutrients and chlorophyll a. *Freshwater Biology* **32**, 113–121.

Dent C.L. & Grimm N.B. (1999) Spatial heterogeneity of stream water nutrient concentrations over successional time. *Ecology* **80**, 2283–2298.

Fisher S.G., Gray L.J., Grimm N.B. & Busch D.E. (1982) Temporal succession in a desert stream ecosystem following flash flooding. *Ecological Monographs* **52**, 93–110.

Fisher S.G., Grimm N.B., Marti E. & Gomez R. (1998) Hierarchy, spatial configuration, and nutrient cycling in a desert stream. *Australian Journal of Ecology* **23**, 41–52.

Grimm N.B. (1993) Implications of climate change for stream communities. In: Kareiva P.M., Kingsolver J.G. & Huey R.B. (eds) *Biotic Interactions and Global Change*. Sinauer Associates Inc., Sunderland, Massachusetts.

Grimm N.B. & Fisher S.G. (1986) Nitrogen limitation in a Sonoran Desert stream. *Journal of the North American Benthological Society* **5**, 2–15.

Grimm N.B. & Fisher S.G. (1989) Stability of periphyton and macroinvertebrates to disturbance by flash floods in a desert stream. *Journal of the North American Benthological Society* **8**, 293–307.

Jeffries M. (1994) Invertebrate communities and turnover in wetland ponds affected by drought. *Freshwater Biology* **32**, 603–612.

Lake P.S., Bayly I.A.E. & Morton D.W. (1989) The phenology of a temporary pond in western Victoria, Australia, with special reference to invertebrate succession. *Archiv für Hydrobiologie* **115**, 171–202.

Matthaei C.D., Peacock K.A. & Townsend C.R. (1999) Patchy surface stone movement during disturbance in a New Zealand stream and its potential significance for the fauna. *Limnology and Oceanography* **44**, 1091–1102.

McCall P.L. & Soster F.M. (1990) Benthos response to disturbance in western Lake Erie – regional faunal surveys. *Canadian Journal of Fisheries and Aquatic Sciences* **47**, 1996–2009.

Peterson C.G. (1996) Mechanisms of lotic microalgal colonisation following space-clearing disturbances acting at different spatial scales. *Oikos* **77**, 417–435.

Rasmussen J.B. & Rowan D.J. (1997) Wave velocity thresholds for fine sediment accumulation in lakes, and their effects on zoobenthic biomass and composition. *Journal of the North American Benthological Society* **16**, 449–465.

Ten Winkel E.H. & Davids C. (1985) Bioturbation by cyprinid fish affecting the food availability for predatory water mites. *Oecologia* **67**, 218–219.

Townsend C.R. (1989) The patch dynamics concept of stream community ecology. *Journal of the North American Benthological Society* **8**, 36–50.

Chapter 9 *What is the impact of predators in freshwater systems?*

9.1 Observing predation

It is a sunny, windless day alongside a derelict canal in southern England. The water is less than a metre deep, and a large pike can be seen hanging motionless alongside a clump of common reeds. Even in the clear water, its mottled green and brown colouration provides near perfect camouflage. To a shoal of small roach working their way along the muddy bottom of the canal, the pike looks like any other piece of woody debris. The roach, unaware of the danger, continue to advance as they pick small invertebrates off the mud.

The pike begins to rotate to face the roach, perhaps alerted to their presence by vibrations that the small fish transmit through the water. The pike's body is soon facing the roach, its turn having been completed with no obvious motion of its body or fins. It now starts to drift towards the roach, an apparently benign log concealing ferocious intent. Roach at the front of the shoal keep advancing, racing each other to the next morsel lying on the mud. The leaders are now less than a metre from the pike. The pike's large eyes focus on a single roach positioned slightly to one side of the shoal. The water erupts in a blur of movement as the pike launches itself at the roach. A swirl of mud hides the fish from view.

It is a minute before the mud settles, revealing the pike once again hanging alone in the water. The roach are nowhere to be seen, the shoal having scattered to nearby beds of aquatic plants or macrophytes. The pike gulps, its gill covers briefly flare open, and a puff of silvery scales glitter in the sunlight before settling to the bed of the canal. The pike will not have to feed again today.

Box 9.1

Defining predation

Predators can be simply defined as animals that either totally or partly consume and harm another animal. However, this very simple and general definition of predation conceals the range of very different types of interactions that might be described as predation. A variety of different classification schemes has been devised with the aim of further describing the variety of interactions that might be described as forms of predation. Whilst no scheme is perfect, the classifications presented below are widely used.

- **True predators** are those animals that attack, kill and eat all or part of their prey. Over their lifetime, a true predator will typically kill many prey animals.
- **Parasitoids** are insects (mostly Hymenoptera and some Diptera) in which the larvae develop and grow by slowly consuming a single host animal (usually the larvae of another insect, or a spider).

Either a single parasitoid or a group of parasitoids develops from the one host. Consequently, each parasitoid kills a maximum of only one prey individual over their complete lifecycle.

- **Parasites** are animals which live in close association and gain sustenance from their host. However, unlike true predators and parasitoids, they do not usually kill their host. They may however severely debilitate their host.
- **Herbivores** are animals that eat plants. The term herbivory includes a great diversity of interactions. Animals, that consume whole plants, are comparable to true predators, whereas animals, that consume only part of a plant, are behaving more like parasites.

This chapter focuses on the impact of predatory fish which are behaving as true predators. The variety of ways in which we can define predators does, however, give some indication of the range of impacts predation can have.

Predation (Box 9.1) is perhaps the most dramatic of biological interactions, and attacks by predatory fish on smaller fish species can be readily observed by anyone with the patience to sit by a clear lake or stream. Our direct observations and intuition suggest that attacks by pike on roach are likely to have a significant impact on the ecology of roach. Roach could become locally extinct if the rate of predation by pike exceeded the number of roach spawned in the canal each year. The roach are also likely to avoid high-risk open water areas of the canal and increase their use of macrophyte beds after a pike attack. However, macrophyte beds may be a habitat that offers few feeding opportunities for a fish better suited to open water habitats. In the long term, the growth and reproductive success of roach in the canal may suffer. Elimination of the roach, or even just restriction of the roach to a particular habitat, may also have important consequences for the invertebrates that the roach feed upon.

9.2 Does fish predation structure aquatic communities?

Quantifying the impact of predation on a prey community requires consideration of a number of factors including definition of the spatial and temporal scales of study. Defining the spatial and temporal scale of the study is often the first step in any ecological investigation (see Chapter 2). Predation can be studied at several different levels of spatial and temporal scale. At the landscape scale, the biota of whole lakes or streams that have or lack a particular predator can be compared. At smaller spatial scales, we can compare micro-habitats within a lake or stream, perhaps comparing areas that might offer a potential refuge from the predator, such as macro-phyte beds, with non-refuge areas such as bare substrates. In many fresh-water systems the presence of a predator, such as a species of fish, may vary over daily or seasonal cycles. The abundance and distribution of potential prey when the fish are present or active can be compared with periods when predators are absent or inactive.

9.3 Does land use or predation determine the distribution of galaxiid fish?

9.3.1 Describing the pattern

Prior to the arrival of Europeans in New Zealand, upland streams were dominated by several species of fish in the family Galaxiidae (Fig. 9.1a). Galaxiids are widely distributed throughout New Zealand, southern Australia and southern South America. However, their abun-dance in New Zealand and Australia appears to have declined markedly since the advent of European colonization. Habitat degradation caused by changes in land use and the introduction of various species of trout for sport-fishing (Fig. 9.1b) from North America and Europe have been implicated as likely causes of their decline.

Determining the relationships between galaxiids, land use and trout distribution requires a study conducted at a relatively large spatial scale, i.e. across a number of streams that drain catchments with varying types of land use, and support a mix of either one or both fish species. Townsend and Crowl (1991) employed a study design that incorporated these fea-tures to determine the primary cause of declines in galaxiid abundance in the Taieri River catchment in southern New Zealand. A total of eight streams draining catchments dominated by one of four different types of land use, tussock grassland, agricultural pasture, pine plantation and native woodland, was selected (Table 9.1 & Fig. 9.2). Within each of the selected catchments, the main stem and three tributary streams were selected for sampling. Within each of these main stem and tributary streams, three pool/riffle sequences were selected. The advantage of this type of study design is that it allows variation in fish abundance to be compared both across catchments and within individual catchments. The composition of fish species at each sampling site was determined by electrofishing. A number of additional habitat variables, including eleva-tion, position in relation to waterfalls, site width and depth, flow velocity and turbidity, was also determined for each survey site.

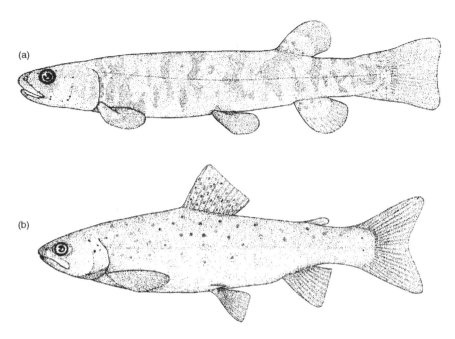

Fig. 9.1 Illustration of (a) common river galaxias (*Galaxias vulgaris*) and (b) brown trout (*Salmo trutta*). (After McDowall 1990 with permission of Reed Publishing (N.Z.) Ltd.)

Table 9.1 Means (and standard deviations) for selected physical properties of each catchment included in the survey of fish distribution by Townsend and Crowl (1991). (After Townsend & Crowl 1991.)

Catchment	Land use	n	Width (m)	Depth (cm)	Elevation (m)	Flow (m/s)	SS(g/L)
					Physical variables		
Silver Stream	Bush	24	3.3 (1.6)	19.5 (10.0)	158 (67)	0.22 (0.25)	1.4 (0.6)
Traquair Burn	Bush	24	3.4 (2.8)	19.1 (11.1)	140 (47)	0.16 (0.15)	3.6 (1.0)
Kye Burn	Tussock	24	3.1 (1.4)	24.9 (17.6)	713 (73)	0.90 (0.91)	1.3 (0.6)
Sutton Stream	Tussock	30	2.2 (1.2)	18.8 (14.0)	807 (141)	0.28 (0.21)	2.1 (2.2)
Big Stream	Pine	24	2.3 (1.0)	16.9 (9.0)	192 (91)	0.27 (0.18)	1.3 (0.6)
Sheperds Creek	Pine	18	2.3 (1.1)	17.1 (11.7)	436 (38)	0.32 (0.24)	1.0 (0.6)
Lee Stream	Agriculture	30	2.3 (1.5)	20.1 (9.2)	416 (75)	0.40 (0.21)	6.0 (1.3)
Nenthorn	Agriculture	24	2.1 (1.0)	13.0 (9.0)	355 (43)	0.11 (0.10)	3.2 (0.7)

SS, suspended solids.

Prior to the study, it was known that common river galaxias (*Galaxias* spp.) and brown trout (*Salmo trutta*) were widely distributed across the Taieri River catchment, although the precise distribution of the two fish species in relation to each other was not known. If land use was the primary cause of the decline in the range of galaxiids, it might be

127

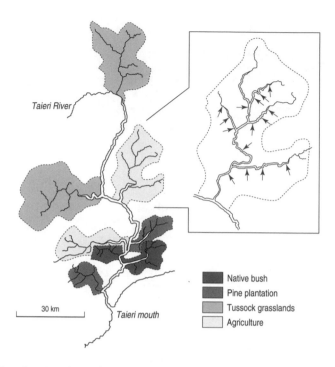

Fig. 9.2 Map showing the eight Taieri River catchments included in the survey of fish distribution and their land usage. Inset illustrates sampling sites within one catchment. (After Townsend & Crowl 1991 with permission of Oikos.)

expected that native galaxiids would remain abundant in catchments dominated by indigenous vegetation, i.e. tussock grasslands and native bush, whereas introduced trout would dominate in catchments dominated by introduced plant species, i.e. pasture grasslands and pine plantations. Alternatively, if brown trout were the primary cause of declines in galaxiid distribution, then an absence of galaxiids in the presence of trout, irrespective of land use, would be expected.

The key finding of the study was that the distribution of galaxiids and trout were virtually non-overlapping (Fig. 9.3). Out of the 198 sites surveyed, 69 contained only common galaxias, and 63 contained only brown trout. Only nine sites, all in shallow, braided sections of stream, contained both common galaxias and trout. No fish were recorded at 54 of the sites surveyed. Sites containing only common galaxias were typically upstream of 3-m waterfalls, and were at high elevations. Sites typically containing only trout were below 3-m waterfalls, and were at lower elevations compared to the galaxias-only sites. The distribution of galaxiids was effectively limited to relatively short headwater sections of stream, whilst the trout occupied the larger, deeper downstream reaches.

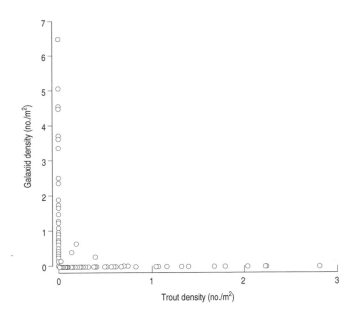

Fig. 9.3 Relationship between trout and galaxiid densities in the Taieri River catchment. (After Townsend & Crowl 1991 with permission of Oikos.)

There was no evidence to suggest that catchment land use was closely related to the fish species present, with both galaxiids and trout occurring across all land use types.

9.3.2 Interpreting patterns of trout and galaxiid distribution

It is not difficult to envisage a range of possible mechanisms that could generate the pattern of galaxiid–trout distribution observed by Townsend and Crowl. Galaxiids may simply prefer to live in streams above water-falls, whereas trout may prefer to live downstream of waterfalls. Such habitat preferences could, in some way, be related to factors such as food availability or reproduction. Alternatively, either trout or galaxiids may be excluding the other species from sections of stream that they currently occupy through either predation or competition. Determining which mechanism may be responsible could be achieved through experimentation, for example introducing trout into sections currently occupied by galaxiids. Such experimental work could, however, run into significant ethical problems. Such an experiment would mean introducing trout into the habitat of a potentially threatened native fish. Experimentation may, however, prove unnecessary if there is convincing additional evidence that leaves us with little doubt as to the mechanism generating the pattern.

Townsend and Crowl (1991) argued that their study demonstrated, beyond reasonable doubt, that predation by trout is the key determinant of galaxiid distribution across the Taieri River catchment. They relied on several lines of evidence to support their argument. Trout are efficient piscivores, and have been associated with the elimination of galaxiids from other areas where they have been introduced, e.g. Australia (see Closs & Lake 1996). Prior to the introduction of trout in the 1870s, records also suggest that galaxiids were widespread throughout the Taieri River system, occupying many stream reaches now occupied only by trout. Such evidence suggests that galaxiids are quite capable of surviving in streams downstream of waterfalls, but for the presence of trout. Galaxiids only survive upstream of 3-m waterfalls due to the inability of trout to cross such barriers. The only factor that appears to have changed consistently across sites now occupied only by trout, is the presence of the trout themselves. The two species only coexist now in a few shallow, braided stream sections where galaxiids could avoid the few trout that were present.

9.3.3 Testing hypotheses: have trout had a wider impact on New Zealand stream communities?

Townsend and Crowl's (1991) study provided convincing evidence of an interaction between galaxiids and trout. Like any good descriptive study, it has also provided the material from which testable hypotheses can be generated. An obvious question that arises from Townsend and Crowl's study is, what happens to stream communities when an exotic predatory fish replaces a native predatory fish? Given that the New Zealand stream fauna evolved in the presence of galaxiids, it seems reasonable to expect that they might have effective anti-predator responses against such a predator. However trout are a novel predator in New Zealand systems, and predator avoidance strategies that work against galaxiids may not be so effective against trout.

Flecker and Townsend (1994) examined the effect of either no fish, galaxiids or trout, on invertebrate communities in New Zealand streams. The obvious way to examine the impact of fish predation on invertebrates in streams is to restrain fish in one area using fences or cages. Experimentally caging fish in streams, with the aim of determining the impact of their predation is, however, fraught with problems. Fish frequently escape, the mesh from which any cages are constructed tends to clog with water-borne debris, considerable variability between cages can render detection of treatment effects difficult, and floods often wash away experiments that run for any length of time. Flecker and Townsend (1994) circumvented some of these problems by constructing 12 artificial stream channels and placing them in a real stream, and then had either no fish, galaxiids or trout treatments run in selected channels for a period of time (Fig. 9.4).

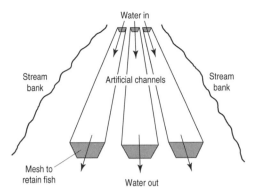

Fig. 9.4 The artificial stream channels used by Flecker and Townsend (1994) to study the impact of fish predation on stream invertebrate communities.

The advantage of such channels is that, with the exception of the manipulated variable, relative uniformity of conditions across all replicates can be achieved. Such uniformity is rarely possible in the highly variable environment of a real stream. The disadvantage of artificial stream channels is that they are not real streams, hence some degree of realism is inevitably lost.

The channels were constructed to replicate real streams as closely as possible. The bed of each channel was layered with gravel and cobbles. Water for the channels was drawn from the stream, allowing invertebrates and algae to colonize naturally. After a 10-day colonization period, four replicates of three treatments, each consisting of either eight galaxiids, eight juvenile trout or no fish, were set up. The experiment was allowed to run for a further 10–11 days. At the conclusion of the experiment, all invertebrates within each channel were collected, and an algal sample was scraped from cobbles within each replicate channel.

Flecker and Townsend (1994) observed that the density of invertebrates in channels containing fish was reduced relative to the no-fish channels (Fig. 9.5a), suggesting that fish presence did have an effect. However, the final invertebrate densities were much lower in the channels containing trout (2685 $\pm$ 404 invertebrates/m^2) compared with galaxiid channels (3952 $\pm$ 457 invertebrates/m^2). A number of individual invertebrate species, such as the grazing mayfly *Deleatidium*, showed marked differences in abundance (no fish: 2952.2 $\pm$ 255; galaxias: 2528.7 $\pm$ 392.1; trout: 1635.7 $\pm$ 265.2), although no invertebrate species were eliminated altogether. The complex substrate within each channel probably provided some protection from each species of fish. Changes in invertebrate density in relation to the fish were most probably due to predation, although changing rates of invertebrate emigration could have also produced an identical result. The number of invertebrates drifting along with the stream current may, in part, be related to fish predation (Malmqvist 1988). Significantly, the introduced predator (trout) had a greater impact on the stream invertebrate communities than the native predator (galaxiids) suggesting that coevolution between predator and prey may ameliorate the impact of predation to some degree.

The changes in the status of fish in the channels also resulted in a change in algal standing crop (as measured by chlorophyll a). Channels containing no fish contained the least algae, whilst the channels containing trout, supported the most (Fig. 9.5b). Changes in algal standing crop were presumably a result of reduced rates of invertebrate grazing. Reduction in invertebrate grazing rates could have been a consequence of either reduced invertebrate abundance, reduced invertebrate activity or a combination of both. Grazing activity did appear to be reduced in the presence of fish, and in the presence of trout in particular. Distinct grazing 'scars' in algae growing on the channel walls were left by grazing invertebrates. The width of the bands was greatest in the no-fish channels, and narrowest in the trout channels (Fig. 9.6). This suggests that mayflies tended to avoid the risk of exposure to visual-feeding fish by remaining close to the shelter of the rock substrates.

The results of Flecker and Townsend's (1994) channel experiment provide convincing evidence that changes in fish density and species can alter the structure of invertebrate and algal communities in artificial stream channels. Given that the channels were placed in real streams, contained similar substrates, and supported invertebrate and algal communities drawn from the real stream, it is also likely that trout or galaxiids are having a comparable effect in real streams. If so, we would expect streams containing trout to support the highest levels of algal standing crop. Huryn (1998) tested this hypothesis by measuring algal standing crop (as measured by chlorophyll a concentration extracted

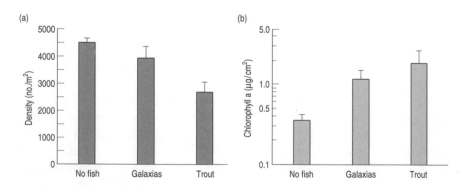

Fig. 9.5 Total invertebrate density (a) and algal standing crop (b) from three fish treatments in experimental channels (means +1 standard error: $n = 4$). (After Flecker & Townsend 1994 with permission of the Ecological Society of America.)

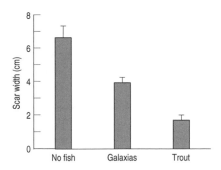

Fig. 9.6 Width of algal grazing scars from three experimental treatments (means +1 standard error: $n = 4$). (After Flecker & Townsend 1994 with permission of the Ecological Society of America.)

from biofilm growing on rocks) in a single trout and a single galaxiid stream at monthly intervals over 12 months. As predicted, over the entire study period, algal standing crop was higher in the trout stream compared to the galaxiid stream (Fig. 9.7). Changes in the patterns of mayfly feeding in trout or galaxiid streams that could contribute to the predicted patterns have been described (McIntosh & Townsend 1996).

9.4 Conclusion: does predation structure freshwater communities?

A number of excellent descriptive and experimental studies of the impact of predation in rivers and lakes have been completed. The patterns

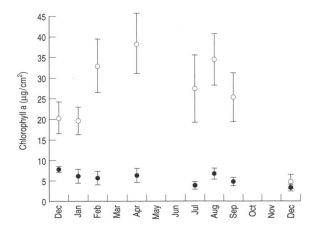

Fig. 9.7 Mean concentrations of chlorophyll a in biofilm in Stony (galaxiids present, open circles) and Sutton (trout present, closed circles) streams (means $\pm$ 1 standard error). (After Huryn 1998 with permission from Springer-Verlag.)

emerging from these additional studies appear to correspond closely with the results of previously described studies. That is, predatory fish can have a major impact on the structure of aquatic communities. Those impacts may be due to the direct impact of predation, or altered behaviour of their prey (McIntosh & Peckarsky 1996, McIntosh & Townsend 1996). Predation by fish in streams tends to reduce the abundance of certain vulnerable species. However, local extinction of any species is rare due to the protection provided by complex stream substrates and macrophytes (see reviews and studies by Power 1992, Dudgeon 1996).

A similar pattern of dynamics appears to be emerging from studies of predation in lakes, although actual local extinction of vulnerable prey species seems to be more common in pelagic habitats compared with benthic habitats (He *et al.* 1994, Beklioglu & Moss 1996, see also Chapter 12). Presumably the complexity of benthic habitats protects benthic invertebrates to some degree from the impact of fish predation. Even in the pelagic zone, potentially vulnerable prey may coexist with predators by migrating vertically or horizontally through the water column to avoid encounters with predators (Lauridsen & Buenk 1996). Local extinction appears to be rare in structurally complex benthic habitats, presumably due to the abundant refugia afforded by such habitats (see reviews and studies by Diehl 1995, Diehl & Eklov 1995). Considerable scope for novel research exists in the areas of how prey behaviour mediates the impact of predation, and how predation impacts on prey populations from generation to generation. There is also relatively little known regarding the impact of invertebrate predators on their prey in comparison with the considerable literature on the effects of fish predation.

9.5 Further reading

Beklioglu M. & Moss B. (1996) Mesocosm experiments on the interaction of sediment influence, fish predation and aquatic plants with the structure of phytoplankton and zooplankton communities. *Freshwater Biology* **36**, 315–325.

Closs G.P. & Lake P.S. (1996) Drought, differential mortality and the coexistence of a native fish (*Galaxias olidus*) and trout (*Salmo trutta*) in an intermittent stream. *Environmental Biology of Fishes* **47**, 17–26.

Diehl S. (1995) Direct and indirect effects of omnivory in a littoral lake community. *Ecology* **76**, 1727–1740.

Diehl S. & Eklov P. (1995) Effects of piscivore-mediated habitat use on resources, diet, and growth of perch. *Ecology* **76**, 1712–1726.

Dudgeon D. (1996) The influence of refugia on predation impacts in a Hong Kong stream. *Archiv für Hydrobiologie* **138**, 145–159.

Flecker A.S. & Townsend C.R. (1994) Community-wide consequences of trout introduction in New Zealand streams. *Ecological Applications* **4**, 798–807.

He X., Scheurell M.D., Soranno P.A. & Wright R.A. (1994) Recurrent response patterns of a zooplankton community to whole-lake fish manipulation. *Freshwater Biology* **32**, 61–72.

Huryn A.D. (1998) Ecosystem-level evidence for top-down and bottom-up control of production in a grassland stream system. *Oecologia* **115**, 173–183.

Lauridsen T.L. & Buenk I. (1996) Diel changes in the horizontal distribution of zooplankton in the littoral zone of two shallow eutrophic lakes. *Archiv für Hydrobiologie* **137**, 161–176.

Malmqvist B. (1988) Downstream drift in Madeiran levadas: tests of hypotheses relating to the influence of predators on the drift of insects. *Aquatic Insects* **10**, 141–152.

McDowall R.M. (1990) *New Zealand Freshwater Fishes: a Natural History and Guide.* Heinemann Reed, Auckland.

McIntosh A.R. & Peckarsky B.L. (1996) Differential behavioural responses of mayflies from streams with and without fish to trout odour. *Freshwater Biology* **35**, 141–148.

McIntosh A.R. & Townsend C.R. (1996) Interactions between fish, grazing invertebrates and algae in a New Zealand stream: a trophic cascade mediated by fish-induced changes to grazer behaviour? *Oecologia* **108**, 174–181.

Power M.E. (1992) Habitat heterogeneity and the functional significance of fish in river food webs. *Ecology* **73**, 1675–1688.

Townsend C.R. & Crowl T.A. (1991) Fragmented population structure in a native New Zealand fish: an effect of introduced brown trout? *Oikos* **61**, 347–354.

Part 3 *Applied Freshwater Ecology*

In Part 2, we examined approaches that could be used to explore four fundamental ecological questions. In the third and final part of this book, we examine the approaches that might be used to examine four applied ecological questions in freshwater systems: the regulation of water regimes, pollution, biomanipulation of food webs to improve water quality, and managing the impact of introduced species. In our view, the distinction between fundamental and applied research is somewhat arbitrary. Whereas curiosity may be the primary motivation driving a researcher investigating a fundamental ecological question, such research frequently has applied applications. Conversely, whilst an applied research project may be primarily motivated by the need to address a specific management problem, such research frequently both provides and depends upon basic insights into the functioning of ecosystems in general. Curiosity is also a key motivator in much applied research, given that the interaction between humans and their environment can provide an abundance of interesting questions to address for the imaginative scientist.

Irrespective of whether research is driven purely by curiosity or the need to address an environmental problem, the quality of the science determines its potential value. Well-designed research provides information in which we can have a high degree of confidence. That high-quality information can be used to improve our understanding of how the world around us works, or can be used directly to address environmental problems. In contrast, science that attempts to address either fundamental or applied questions poorly only wastes money and resources. In an applied context, it can even be dangerous if inaccurate results are used to manage endangered ecosystems, threatened species or people's lives.

Chapter 10 *What are the ecological effects of changing a water regime?*

10.1 The Aral Sea disaster – the result of changes in water regime

Since 1960, the water regime of the Aral Sea in central Asia has been dramatically altered by the progressive diversion of its two major inflowing rivers – the Syr-Dar'ya and Amu-Dar'ya – for irrigation. From 1960 to 1990, the water level in the lake has fallen from 53.4 m above sealevel to 38.6 m above sealevel, the lake's area has declined from 68 000 to 33 500 km^2 and its volume has been reduced by over two-thirds to 310 km^3. Fishing villages are now stranded many kilometres from the present shore. But this is now of little relevance to the people who once fished for a living because the changes in the lake's water regime has meant that the annual fish-catch, over 40 000 tons before the 1960s, had declined to zero by the 1980s.

Why did a change in the water regime of the Aral Sea result in what is now considered one of the globe's most serious ecological disasters? Prior to the 1960s, the Aral Sea was slightly saline (c. 10 g/L) but supported an essentially freshwater fauna of about 20 fish species, over 350 species of aquatic invertebrates and 12 species of higher water plants. With the diversion of water after 1960, three factors began to interact: falling water levels, salinization, and reduction of the rate of input of plant nutrients. Salinization and reduced nutrient inputs greatly impacted the phytoplankton and water plants. By the mid-1970s, average salinity was greater than 14 g/L and phytoplankton biomass had declined up to five-fold. Falling water levels also immediately impacted emergent water plants that rapidly disappeared from the shallow areas and deltas of the two dwindling rivers.

There is a bizarre twist to this story – but one typical of many human impacts on wetlands because seldom is the influence due to a single effect. In addition to the changes in water regime, there had been several introductions of exotic fishes and invertebrates since 1927. The first effect of the introduced fishes was to increase predation pressure on zooplankton and benthic invertebrates so that some taxa were virtually eliminated and benthic biomass declined three-fold. The large, original species of zooplankton were quickly consumed and soon only small taxa, mainly rotifers, remained. Further changes resulted from the increase in salinity. The first signs were seen in 1971 when the average salinity reached 12 g/L, and by the mid-1970s all freshwater zooplankton appeared extinct. However, exotic species tolerant of a wide range in salinity (euryhaline) proliferated and zooplankton biomass started to rise because the planktivorous fish were being eradicated by the rising salinity. As water levels fell further and salinity exceeded 23 g/L during the late 1980s, abundance and diversity of the zooplankton fell sharply. Rising salinity also changed the composition of the benthic fauna – worms and insect larvae disappeared while euryhaline species, including an introduced mussel (*Abra ovata*) dominated.

Falling water levels and rising salinity impacted fish from the mid-1960s. By the end of the 1960s, the fall in water level had reduced the area of the shallow spawning grounds by 80%. Fish were also affected by the construction of weirs and dams on the lower reaches of the Syr-Dar'ya and Amu-Dar'ya. In the early 1970s, adult fish were showing signs of stress from high salinity with growth rates declining. Mortality rates also rose sharply and there were numerous reports of abnormally shaped fish. By the mid-1970s, most natural reproduction had ceased and by the late 1970s, few fish larvae could be found. All fish, introduced and native, were practically extinct in 1990 except near the inputs of the rivers (Aladin & Williams 1993, Stone 1999).

10.2   Introduction

The case of the Aral Sea provides a dramatic illustration of the catastrophic changes that may occur when the water regime of a waterbody is altered beyond the range of conditions under which the natural ecosystem developed. The integrity of the Aral Sea ecosystem collapsed as the natural water regime altered. Declining inflows reduced water quality by increasing salinity. Energy sources declined as falling nutrient availability led to the collapse of primary production and physical habitats altered as the lake became shallower. These changes altered the dynamics of biotic interactions as various fish and zooplankton populations increased and decreased with the changing conditions. As the environment became increasingly extreme relative to the conditions present prior to regulation of the rivers flowing into the Aral Sea, the native fauna declined in diversity and abundance to the point where they virtually disappeared.

10.3 Determining the impacts of an altered water regime in rivers

Rivers and streams are characterized by variation in the flow of water along them (Poff *et al.* 1997). Ideally, if we want to determine the impact of altered water regime, we would collect data from the system prior to any change in the water regime occurred. This **before**-impact data collection would ideally occur over at least several years so natural seasonal and year-to-year variation was quantified (see Chapter 5). Our sampling regime would ensure that the critical features of a river's water regime were measured (Box 10.1), along with likely biological responses.

Box 10.1

Poff *et al.* (1997) summarized and reviewed the factors influencing the critical components of a river's flow or water regime in an excellent paper published in the journal *BioScience*. The five key components of a flow or water regime are as follows.

The **magnitude** of the discharge at any particular point in time: this simply refers to the amount of water flowing past a particular point over any particular unit of time. In large rivers, a small increase in magnitude may result in a negligible change in water levels. However, in steep, small streams, even a relatively small increase in magnitude may result in a major change in water levels and average water velocities.

The **frequency** of occurrence refers to how often a flow of a particular magnitude will occur. Frequency of occurrence is often expressed as a recurrence interval, for example a one in 100 year flood. Such a flood could be expected to occur, on average, every 100 years. Such a probability can be estimated using previously recorded hydrological data. Of course, such an estimate is only a statement of the probability of a particular event occurring through time. One in 100 year floods can occur in consecutive years. If such events become more frequent (perhaps due to climate change), then recalculation of flood intervals will be required, and the recurrence interval for what would have previously been regarded as a one in 100 year flood would be reduced.

The **duration** is the length of time over which a flow of a certain magnitude is maintained. The duration of the most extreme events (e.g. peak flood levels) may only last a few minutes or hours. However, the ecological consequences of such events can have long-lasting effects on community structure (see Chapters 8 and 10).

The **timing** or **predictability** refers to the distribution of particular events through time. Rivers receiving seasonal rainflow may have highly predictable periods of high and low flows. In contrast, patterns of flow may be very unpredictable in arid zone streams (see Chapter 5).

Rate of change or **flashiness** refers to how quickly flow changes from one magnitude to another. Flashy streams have high rates of change. Typically, such streams have relatively small catchments that are subject to short periods of intense rainfall, for example severe thunderstorms. In contrast, streams fed from groundwater springs can have very stable flow regimes.

These five critical components of a flow regime influence water quality, energy sources, the physical structure and biotic interactions. All of these factors interact with each other. Hence, a change to one component of a flow regime can trigger a cascade of interrelated changes in a river ecosystem. Determining precise cause and effect relationships, particularly in large river or lake systems where multiple changes across a catchment may have occurred, can be virtually impossible.

We would continue to collect data for many years **after** the water regime was altered given that the impact of an altered water regime on the river's biota may take many years to develop fully. Further, we would conduct the same sampling regime on a similar **control** river system in which the water regime remained in a natural state. Sampling in this control system would allow us to separate natural variation from the effects of the altered water regime in the **impacted** system. Such a study design is known as a **before–after–control–impact** (BACI) design.

Unfortunately, opportunities to conduct such well-designed BACI studies are rare. In the case of the Aral Sea and its inflowing rivers, alterations to the water regime began long before anyone ever thought of measuring anything in detail. Further, ecosystem responses to the altered water regime of rivers flowing into the Aral Sea developed over decades. Documenting biotic responses to an altered water regime that develop within a month or perhaps a year is realistic, falling well within the length of time research projects typically run. However, following changes that unfold over a period of time that is perhaps longer than the lifespan of the researchers involved is obviously a more difficult proposition. Lastly, there is only one Aral Sea, hence it is impossible to compare directly between an impacted and a non-impacted Aral Sea.

The case of the Aral Sea highlights how difficult it can be to study relationships between the water regime and an ecosystem response, particularly where we are dealing with large systems. However, whilst it may not always be possible to use the ideal study design to determine the impact of an altered water regime, the case of the Aral Sea does suggest that we can logically infer relationships between water regime and community structure provided we have reasonably accurate long-term measurements of the water regime and key aspects of the community structure. Fortunately, continuous river level records have been kept for many rivers, usually where such information is required for flood control, irrigation, navigation or hydroelectric generation. In some cases, these records may extend back for over 100 years. If accurate records of community structure over a corresponding time period can be accessed, then the potential for a detailed study of relationships between water regime and community structure may exist.

10.3.1 What is the impact of water regime on riparian vegetation?

The plains cottonwood (*Populus deltoides* subsp. *monilifera*) is an important riparian tree across semi-arid parts of North America. River communities are often (although not always!) strongly influenced by the structure of the surrounding riparian vegetation (see Chapter 5) hence changes in the abundance of the dominant riparian species can result in significant changes to the aquatic communities that live alongside. Germination of the seeds of plains cottonwood is usually restricted to bare moist sites. Such conditions are generated by variation in water levels. Vegetative reproduction by root sprouting is relatively uncommon, hence the tree only grows in areas where seeds are able to germinate and then survive to maturity. Whilst germination may occur in any bare, moist sediment, long-term survival will only occur in those areas protected from floods or ice-scour (caused as the winter ice pack breaks up and drifts downstream). A question that confronts river managers is: what features of a river's morphology and water regime are associated with successful cottonwood germination and establishment?

Along the Missouri River, Montana, and other rivers of the region, areas suitable for plains cottonwood germination and subsequent establishment occur along sections of river channel subject to either channel narrowing, channel meandering or flood deposition of seeds (Scott *et al.* 1996). Channel narrowing occurs when a river retreats from an area of former riverbed, perhaps due to down cutting of a channel, diversion of water down an alternative channel or reduction of flow due to upstream damming. Cottonwood trees establish on the bare moist sediment exposed by the falling water levels. Stands that form as a result of channel narrowing are relatively low on the bank and often of variable age, particularly if water levels fluctuated over time. Channel meandering occurs as sediment on the outer bank of a river bend is eroded and then deposited on the inside bank of a downstream bend. Cottonwood seedlings establish on moist, recently deposited sediment. Juvenile trees can then develop and mature if they are protected from floods and ice-scour by subsequent sediment deposition. Cottonwood stands produced as a result of channel meander typically form relatively small, crescent-shaped stands of even-aged trees scattered across the floodplain. Establishment as a result of flood disturbance occurs when floods create areas of bare moist soil above the level at which future smaller, more regular floods occur. Stands of trees that establish after flooding are typically even-aged (all germinating after a single flood event) and the establishment point is high relative to the surface of the channel.

Scott *et al.* (1997) used the conceptual model of cottonwood establishment outlined above to determine which mechanism of cottonwood establishment dominated along a reach of the Missouri River, Montana. The model described previously makes specific predictions with respect to the likely age structure and spatial distribution of cottonwood that is likely to occur in association with the three establishment mechanisms (Table 10.1). Hence, in order to determine which mechanism dominated cottonwood establishment along the Missouri River reach studied by Scott *et al.* (1997), they had to determine (i) the age structure of the trees in the area of interest; (ii) the vertical distribution of the trees above

Table 10.1 Fluvial processes producing sites suitable for cottonwood establishment. (After Scott *et al*. 1997 with permission of the Ecological Society of America.)

Fluvial process	Flow	Cottonwood distribution across floodplain
Channel narrowing	One to several years of flow	Spatial distribution variable; usually not even-aged stands; establishment surface at low elevation of former channel bed
Channel meandering	Frequent moderate flows	Moderate number of even-aged stands arranged in crescent-shaped bands; establishment surface near channel bed elevation
Flood deposition	Infrequent high flows	Small number of linear, even-aged stands; establishment surface well above channel bed elevation

the river channel; and (iii) relate age structure and spatial distribution to hydrological records of river flow dating, in this case, back to 1891. In addition, examination of the geomorphology of the study reach suggested little evidence of channel meandering, hence cottonwood establishment was most likely to be related to either channel narrowing or flood deposition.

Scott *et al*. (1997) determined the age of cottonwood trees along the Missouri River reach of interest by examining growth rings present in the oldest wood they could extract from trees growing along their study reach. The elevation of each tree was also measured relative to the lowest extent of perennial emergent plants in the river channel.

The cottonwood trees sampled along the Missouri River revealed a mixed age structure, a pattern that could be consistent with establishment following channel narrowing or flood deposition. However, comparison of the age distribution with the river hydrograph revealed that age of establishment of trees older than seedlings was strongly associated with floods exceeding $1400\,\mathrm{m^3/s}$ (Fig. 10.1). Of the 64 trees sampled, 35 (55%) began growing in the 2 years following a flood year. However, seedlings were observed growing on the bare moist surfaces near the water's edge (Fig. 10.1) indicating that a supply of seeds capable of germinating on suitable surfaces was available even in non-flood years.

Comparison of the year of establishment of any particular tree and its elevation above the present river channel revealed that older trees were generally located at higher elevations, whereas younger trees were more frequently encountered at lower elevations (Fig. 10.2). This pattern of

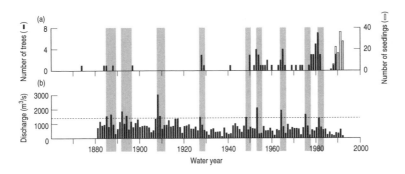

Fig. 10.1 (a) Number of sampled cottonwood trees and seedlings established along the Missouri River, Montana. Solid bars represent tree, sapling and pole size classes. Open bars represent seedling size class (note different scale). (b) Maximum daily discharge for each water year at Fort Benton, Montana. The dashed line indicates a discharge of $1400\,\text{m}^3/\text{s}$. Shaded vertical sections are years with maximum discharge $> 1400\,\text{m}^3/\text{s}$ and the two following years. (After Scott *et al*. 1997 with permission of the Ecological Society of America.)

spatial distribution is again consistent with establishment through channel narrowing or down cutting, with the youngest trees establishing close to the water's edge as river levels fall exposing sediments suitable for germination. However, examination of the river geomorphology and river hydrographs through time revealed no evidence for channel narrowing or down cutting over the period for which the hydrographs were available.

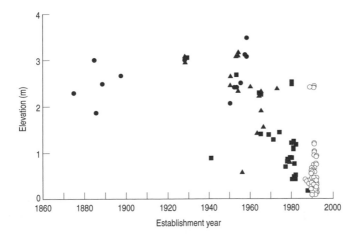

Fig. 10.2 Elevation of establishment surface and year of establishment for sampled cottonwoods. Size classes are represented as trees (●), poles (▲), saplings (■) and seedlings (○). (After Scott *et al*. 1997 with permission of the Ecological Society of America.)

Scott *et al.* (1997) argued that tree survival in the long term declined with decreasing elevation. Most of the seedlings present at the water edge would be subsequently destroyed by either flood or ice-scour in the following year or so. Trees growing at intermediate elevations would also be exposed to more floods and ice, or the full force of the more extreme floods. Long-term survival is likely only for those seedlings that germinated on high, bare, moist locations created by the most extreme floods. Given the association between large floods and cottonwood establishment observed by Scott *et al.* (1997) along the Missouri River reach they studied, it is not hard to envisage how patterns of establishment of cottonwood might be significantly altered if a large dam that eliminated flooding was built upstream of their study reach.

Scott *et al.* (1997) use several lines of independent evidence to support their argument regarding cottonwood establishment along the river reach in which they were interested. No one line of evidence can totally exclude the various models of cottonwood establishment, hence an argument based on limited evidence would be unconvincing. However, when multiple lines of independent evidence are considered together, the most likely mechanism of cottonwood establishment along the reach of interest can be identified with reasonable confidence. Ideally, we would now follow up with an experimental test of their model to confirm the predicted cause and effect relationship. However, an experimental test of their model would require the experimental generation and prevention of major floods along replicate large rivers, followed by years of monitoring. Manipulations on such a scale are virtually impossible due to logistical difficulties and the lack of true replicates (i.e. multiple replicate large rivers to manipulate).

10.3.2 Can we use spatial comparisons to determine the impact of altering water regime?

Examination of the age structure of a population is one approach that we can use to explore the influence of water regime on aquatic communities. However, it requires long-lived organisms that can be accurately aged and hydrological records that extend over the period of interest. Clearly, such a combination of circumstances limits the number of situations we can study given that most organisms are short-lived and hydrological records are lacking for most freshwater systems. Long-term, detailed hydrological records usually exist only for permanent streams/rivers with potential human values as a water supply whereas highly variable intermittent systems are far less likely to be gauged. However, given the variability of intermittent systems, longer term data is required to gain accurate estimates of flow patterns compared with less variable perennial systems.

An alternative approach to studying the impact of water variation is to compare community structure of rivers subject to different water regimes. If the water regime along a particular river reach is altered, we can expect the community along that river to become progressively different compared to a similar reach where the water regime was not altered (assuming that water regime is the driving force). This argument suggests that we can determine the long-term impacts of an altered water regime by spatial comparisons of rivers that are otherwise similar apart from the altered water regime.

Kinsolving and Bain (1993) used this logic to determine whether daily flow fluctuations associated with discharge from a hydroelectric dam altered a fish community relative to a river with a natural water regime. They compared the fish community in a 66-km reach of the heavily regulated Tallapoosa River with a 66-km reach of the Cahaba River, both of which are in Alabama (Fig. 10.3). Each 66-km study reach was divided into 1.6-km reaches, and fish collected using electrofishing from a randomly selected $18\,m^2$ site within each 1.6-km reach.

Two obvious problems confront researchers planning any study that involves spatial comparisons over such a large scale. The effort required to sample adequately the fish community along a 66-km river reach is substantial, hence the number of replicate reaches studied is likely to be low. Indeed, in the study by Kinsolving and Bain (1993), there is no replication at the river scale (see Chapter 1), only a regulated and un-regulated river. Further, as we increase the size of our study reaches, the likelihood that the two reaches will differ also increases, irrespective of whether they are regulated or not. If they do differ significantly, then we will have difficulty in determining whether observed differences in the fish community are due to the impact of river regulation or some other unrelated factor. For the comparison to be useful, the researchers must convince the reader that the fish communities would indeed be relatively similar were it not for the altered water regime on one of the rivers.

In the study by Kinsolving and Bain (1993), the regulated Tallapoosa and unregulated Cahaba River share a number of fundamental features. They are both tributaries of the Alabama River and hence share a common species pool. Both rivers also flow through catchments that are similar in geology, vegetation and gradient. Consequently, the water chemistry in the two rivers is also similar. However, there are significant differences, the most notable being the smaller size of the Cahaba River relative to the Tallapoosa River. Consequently, when presented with differences in the fish community between the two rivers, we have to decide whether the differences are due to the impact of river regulation, or whether they are simply natural differences in the fish communities of each river.

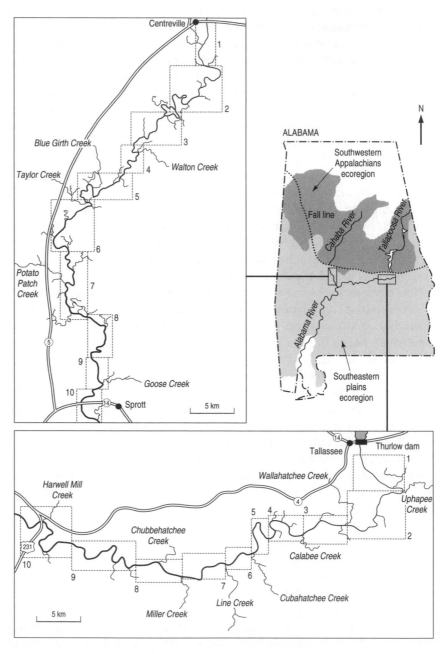

Fig. 10.3 Location of the rivers and reaches used by Kinsolving and Bain (1993). The Alabama map shows the fall line (geological and aquatic biogeographical boundary) and two ecoregions. (After Kinsolving & Bain 1993 with permission of the Ecological Society of America.)

What are the ecological effects of changing a water regime?

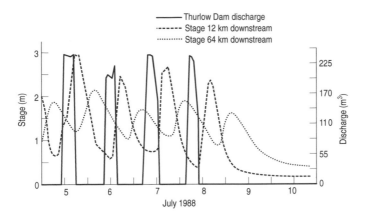

Fig. 10.4 Daily hydrographs (stage or discharge) for a representative 6-day period in the Tallapoosa River showing the fluctuations in water release from the Thurlow hydroelectric dam and water surface elevations at two points downstream. These hydrographs show the effect of discontinuous water releases on downstream water levels and the attenuation of the water level fluctuations with distance downstream. (After Kinsolving & Bain 1993 with permission of the Ecological Society of America.)

The water regime on the Tallapoosa River is altered by the Thurlow Dam, a hydroelectric system 2 km upstream from Kinsolving and Bain's study reach. The Thurlow Dam power station is generally used during periods of peak demand for electricity generation, hence water typically flows through the system for relatively short periods. When no electricity is being generated, no water is released through the dam. As a result, the river just downstream of the dam alternates dramatically between periods of zero flow and lentic conditions to turbulent, fast-flowing conditions. The river levels vary by as much as 3 m between flow and non-flow periods (Fig. 10.4). Tributary streams join the Tallapoosa River downstream from the Thurlow Dam, and as a result the severity of fluctuation in water flow is reduced with increasing distance downstream from the dam. Sixty-four kilometres downstream of the dam, the river continues to flow irrespective of the discharge through the dam and the fluctuation in water level is reduced to approximately 1 m.

If the fish community in the Thurlow Dam was altered as a result of the regulated water regime, Kinsolving and Bain (1993) predicted that the fish community would exhibit some degree of recovery and become increasingly similar to the fish community in the Cahaba River with increasing distance downstream of the Thurlow Dam. Specifically, they predicted that the number of fluvial fish (specialist river fish) species would be reduced immediately downstream of the Thurlow Dam, but would then increase as the water regime approached its natural state with increasing distance downstream from the dam. In contrast, the number of fluvial species was expected to exhibit no trend along the unregulated Cahaba River study reach. The number of generalist fish species was predicted to show no downstream gradient along either river. Generalist fish species were considered to be those species commonly found in both lentic and lotic systems, and hence should be able to cope equally well with the regulated and unregulated water regimes of the Tallapoosa and Cahaba Rivers, respectively.

Kinsolving and Bain (1993) plotted diversity of fluvial and generalist fish species recorded along each study reach on both rivers (Fig. 10.5). As predicted, the number of fluvial species increased with increasing distance downstream from the Thurlow Dam. However, no clear trends in the number of fluvial species were evident along the study reach on the Cahaba River. Also as predicted, no trends in the number of generalist species were evident along either river.

The study of longitudinal patterns of fish distribution by Kinsolving and Bain (1993) clearly has some important limitations, the most obvious being the lack of replication. However, limited replication is a common feature of many large-scale studies (see also Chapter 12). The paired system approach used by Kinsolving and Bain (1993) does allow us to have greater confidence that the longitudinal change in fish community observed in the Tallapoosa River is a consequence of the altered water regime. Longitudinal variation is a common feature of river communities (see Chapter 5), hence the observation of a gradient of change

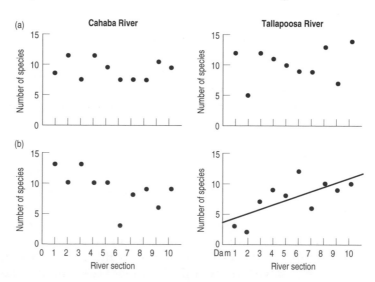

Fig. 10.5 Relations between the number of recorded species and river section for (a) macro-habitat generalists and (b) fluvial specialists in the Cahaba and Tallapoosa Rivers. Only fluvial species showed a significant relationship between the number of species and distance downstream from Thurlow hydroelectric dam. (After Kinsolving & Bain 1993 with permission of the Ecological Society of America.)

downstream from a dam could simply be a natural pattern. However, the absence of any comparable longitudinal gradient in the fish community of the Cahaba River suggests that at least over the length of river reach surveyed in this study, strong natural longitudinal changes in fish community structure are unlikely in a system such as the Tallapoosa River. The observation of similar gradients of invertebrate community recovery downstream of dams (e.g. Voelz & Ward 1991) also suggests that alterations to the water regime can result in significant changes to riverine communities.

10.4 Can we predict the impact of an altered water regime?

Large-scale comparisons of communities affected by different water regimes can tell us whether alteration of the water regime impacts on community structure. However, whilst such descriptive studies provide us with an indication of a change, the precise mechanisms driving changes in the abundance of various species in the community may remain unclear. In the study by Kinsolving and Bain (1993), changes in the fish community with respect to water regime are apparent. However, why the abundance of any particular species of fish has changed is uncertain, particularly given that different fish may respond to the altered water regime in many different ways. For some species, loss of stable flows required for breeding may be critical, whereas others may rely on a food source such as drifting invertebrates that are only present when the stream is flowing. In some cases, a change in the abundance of a species may not be a direct response to an altered water regime, but perhaps a response to a change in the abundance of a predator or competitor that is sensitive to the altered water regime.

10.4.1 How does an altered water regime change a community?

During the droughts that affected northern Californian rivers in 1990–92 and 1994, Wootton et al. (1996) observed marked increases in the abundance of a predator-resistant limnephilid caddisfly *Dicosmoecus gilvipes*. *Dicosmoecus* achieves a degree of resistance to both invertebrate and fish predators by growing to a large size and constructing a heavy predator-resistant case (Fig. 10.6). However, this protection comes at a cost. The large size and heavy case of *Dicosmoecus* restrict it to the river bottom and prevents it from accessing interstitial spaces between rocks. As a result, *Dicosmoecus* is unable to avoid the full force of floods and suffers high mortality during such events. Consistent with the observation that *Dicosmoecus* are unable to resist floods, Wootton et al. (1996) observed a

Fig. 10.6 The heavy protective case in which the larva of *Dicosmoecus* lives. The case is constructed from small stones held together by fine silk. (After Wiggins 1996.)

77% reduction in the abundance of *Dicosmoecus* following a small flood in April 1992.

Wootton *et al.* (1996) hypothesized that the increase in the abundance of *Dicosmoecus* during droughts was due to an absence of mortality associated with flooding and their resistance to predators. Further, they hypothesized that increases in the abundance of *Dicosmoecus* would have consequences for the wider stream community. *Dicosmoecus* grazes algae, hence high abundances of *Dicosmoecus* would divert energy away from other predator-susceptible grazers and invertebrate predators. In the absence of floods, Wootton *et al.* (1996) argued that the configuration of the stream food web would change from one leading from benthic algae through small invertebrate grazers to invertebrate predators and then fish, to a food web that led from benthic algae to the predator-resistant *Dicosmoecus*.

The hypothesis that changes in the abundance of *Dicosmoecus* regulates the abundance of benthic algae is one that lends itself to a small experimental study. Wootton *et al.* tested the hypothesis that high abundances of *Dicosmoecus* would reduce the abundance of benthic algae, thus diverting resources from other invertebrates that were vulnerable to fish predation in artificial stream channels. Twenty-four channels were constructed in the bed of the South Fork River, California. Each channel was 1.56 m long, 1.17 m wide and 0.78 m deep, and had a 5 cm deep gravel substrate. Eight ceramic tiles from which algae could be sampled were placed in each channel. On 24 June 1992, 120 *Dicosmoecus* were introduced to half of the channels. In addition, three juvenile (40–80 mm) steelhead trout (*Onchorynchus mykiss*) were put in half of the channels containing *Dicosmoecus* and half of the channels with no *Dicosmoecus*. It was hypothesized that the addition of the steelhead trout would also alter the community, largely through a reduction in the abundance of predatory invertebrates. The experiment was then allowed to run for 1 month.

As predicted by Wootton *et al.* (1996), high densities of *Dicosmoecus* did reduce benthic algae in the channels (Fig. 10.7) suggesting that energy that would otherwise be available to other invertebrates and ultimately fish was being diverted into a predator-resistant food chain. *Dicosmoecus* also markedly reduced the abundance of sessile invertebrates through a combination of direct predation and disturbance, a result that was not predicted by their original model, and resulted in a further reduction in resources that would ultimately be available to trout. Steelhead trout also had a clear influence on community structure, reducing abundance of predatory invertebrates by 62%, a result which allowed the abundance of sessile invertebrates to actually increase.

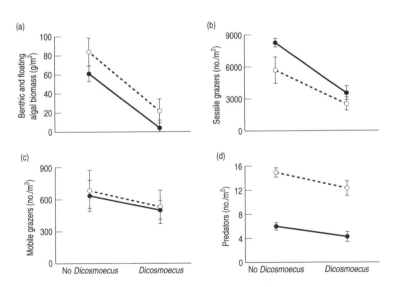

Fig. 10.7 Mean (± 1 standard error) responses of (a) algal grazers, (b) sessile predator-susceptible grazers, (c) mobile predator-susceptible grazers and (d) small predators to the manipulation of the predator-resistant limnephilid caddis grazer *Dicosmoecus gilvipes* and steelhead trout (*Onchorynchus mykiss*). Lines indicate treatments with steelhead present (solid line) and absent (dashed line). (After Wooton *et al.* 1996 with permission from the American Association for the Advancement of Science.)

The results of the channel experiment by Wootton *et al.* (1996) confirmed the possibility that energy from benthic algae can move along two different pathways in the stream food webs they studied. Where *Dicosmoecus* is rare or absent, energy will move from benthic algae through various invertebrate species to steelhead trout. In contrast, where *Dicosmoecus* is abundant, energy moves from benthic algae directly to the predator-resistant *Dicosmoecus*.

Wootton *et al.* (1996) obtained this clear result in the controlled environment of their stream channels and not in the more variable and complex environment of a real stream. To determine whether the results of their channel experiment were applicable to real streams, Wootton *et al.* (1996) compared the food webs of unregulated, regularly flooded rivers, and regulated, rarely flooded rivers. Based on their original observation of the inability of *Dicosmoecus* to resist floods, it was hypothesized that *Dicosmoecus* would be more abundant in the regulated and rarely flooded rivers and rare in the unregulated and regularly flooded rivers, especially during non-drought periods. Surveys of invertebrate communities in a number of regulated and unregulated rivers did indeed find that predator-resistant grazers such as *Dicosmoecus* were more abundant in the regulated rivers (Fig. 10.8). Further, changes to the structure of the food web in the regulated and unregulated rivers were consistent with the patterns observed in the experimental channels.

10.5 Altering the water regime can have complex effects

The results of Wootton *et al.* (1996) have several important implications. From the perspective of those wishing to manage trout populations in certain Californian streams, the results suggest that reductions in the frequency of natural flooding can lead to a reduction in trout abundance. The construction of dams has been implicated in the decline of many salmonid populations, either through the prevention of spawning migrations or the death of fish as they pass through hydroelectric turbines. However, the results of Wootton *et al.* (1996) suggest that other, more subtle changes to community structure and patterns of energy flow can also lead to reductions in the abundance of salmonid fish such as steelhead trout following river regulation.

The study by Wootton *et al.* (1996) also has wider implications. Their study highlights the fact that alteration of a water regime can have complex and often unexpected effects on community structure and function. Simple changes to the amount of water in a system impacts on the multiple elements that make up an ecosystem. Interactions between multiple factors can produce the complex and often unexpected

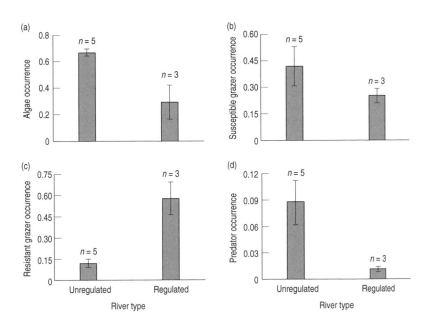

Fig. 10.8 Average ($\pm$ 1 standard error) occurrence during the summer growing season of (a) visibly conspicuous algae, (b) predator-susceptible grazers, (c) predator-resistant grazers and (d) predators in unregulated (with flood disturbance) and regulated rivers (flood disturbance greatly reduced) in northern California. (After Wooton *et al.* 1996 with permission from the American Association for the Advancement of Science.)

dynamics that we observe. The study by Wootton *et al.* (1996) also demonstrates that aspects of large-scale ecological patterns can be effectively studied in small-scale experiments. Their large-scale descriptive study led to the generation of a simple hypothesis regarding the mechanism driving the pattern. Inferences of cause and effect based only on descriptive studies are invariably weak. However, if the mechanism driving the pattern can be demonstrated experimentally, confidence in the interpretation of the patterns observed increases immeasurably. Their study is an excellent demonstration of the logical power of good science.

10.6 Further reading

Aladin N.V. & Williams W.D. (1993) Recent changes in the biota of the Aral Sea, central Asia. *Internationale Vereinigung für Theoretische und Angewandte Limnologie* **25**, 790–792.
Kinsolving A.D. & Bain M.B. (1993) Fish assemblage recovery along a riverine disturbance gradient. *Ecological Applications* **3**, 531–544.

Poff N.L, Allen J.D., Bain M.B. *et al.* (1997) The natural flow regime. *BioScience* **47**, 769–784.

Scott M.L., Friedman J.M. & Auble G.T. (1996) Fluvial process and the establishment of bottomland trees. *Geomorphology* **14**, 327–339.

Scott M.L., Auble G.T. & Friedman J.M. (1997) Flood dependency of cottonwood establishment along the Missouri River, Montana, USA. *Ecological Applications* **7**, 677–690.

Stone R. (1999) Dying seas – Aral Sea. Coming to grips with the Aral Sea's grim legacy. *Science* **284**, 30.

Voelz N. & Ward J.V. (1991) Biotic responses along the recovery gradient of a regulated stream. *Canadian Journal of Fisheries and Aquatic Sciences* **48**, 2477–2490.

Wiggins G.B. (1996) Trichoptera families. In: Merritt R.W. & Cummins K.W. (eds) *Aquatic Insects of North America*, 3rd edn. Kendall/Hunt Publishing Co., Dubuque.

Wootton J.T., Parker M.S. & Power M.E. (1996) Effects of disturbance on river food webs. *Science* **273**, 1558–1561.

Chapter 11   *How do we assess the impact of pollution?*

11.1   The smell of failure and success: the history of pollution in the River Thames

The River Thames drains fertilized farmland before flowing through London with a population of about 10 million. For almost 2000 years, the river has received – and largely absorbed – the wastes of London. But the introduction of the water closet in 1800 and the rapid rise in human population to over 2 million during the subsequent 50 years exceeded the river's natural capacity to degrade organic pollution. By 1850, the river's fisheries were destroyed and Londoners complained bitterly that the water 'stank foully of rotten eggs' (probably hydrogen sulfide). So bad was the stench that the British Parliament, situated on the smelly river's bank, enacted the world's first major sewage diversion scheme, piping the waste downstream for discharge during the ebb tide. This diversion, together with other waste treatment works, temporarily improved the local conditions, and migratory marine fish returned to the river early last century.

However, as London's population increased and new industries began to add toxic effluents to the river, the fishery crashed again. By 1950, only a few eels were recorded in the Thames near London. Not only had dissolved oxygen concentrations plummeted to less than 10%, hydrogen sulfide wafting up from the anoxic mud exposed when water levels were low had blackened the lead-based paints on buildings near the river. The Thames, once the proud centrepiece of England's capital, was a polluted disaster zone. Could the river be repaired? If so, how could the *causes* be addressed?

157

A river restoration programme commenced in the mid-1960s, starting with upgrading sewage treatment to the secondary level (i.e. removing much of the nutrient load) to reduce the oxygen deficit in the river. Newer sewage treatment plants also added oxygen to the effluent to convert toxic ammonia to nitrate, and further reduce oxygen demand from the river. There were also stricter regulations on the discharge of pollutants from industry into the river. The effects were almost immediate. Fish species richness has now increased to over 100 species from the lone eel species in 1950. In 1966, a single salmon was caught in the river but since the early 1990s, a regular run of up to 200 Atlantic salmon per year occurs annually.

11.2 Some background information on water pollution

11.2.1 What is water pollution?

Before we embark on a case study of assessing pollution in an aquatic ecosystem, we need to define pollution carefully. **Pollution** is defined formally as any environmental state or manifestation which is harmful or unpleasant to life, resulting from failure to achieve or maintain control over the chemical, physical or biological consequences of human activities (paraphrased from Collocott & Dobson 1974). The two key issues here are **negative impact on life** and **human fault**. Most of us are familiar with examples of local water pollution, especially in urban areas where stormwater runoff and excessive use of fertilizers lead to a cocktail of nutrients, heavy metals, oils and greases, plastics, industrial and domestic wastes and toxicants, and discarded food packaging. We also know that while these pollutants are generally 'accidental', they have a negative impact on the environment and can be blamed on human activities.

Humans have a long tradition of using rivers as suitable conduits for removing wastes. While human populations were small and non-industrialized, most river ecosystems were able to cope with the biological loads that were dumped into them. As we saw in the above example, the River Thames has been inhabited for over 2000 years but the really serious impacts of pollution such as widespread fish kills were not evident until a few hundred years ago. However, there was an awareness of pollution long before then – in the 13th century, laws were passed prohibiting charcoal burners in Middlesex from washing their charcoal in the Thames because of the downstream effects. One key point is that not only has the capacity of the river to absorb pollution been exceeded but that the **types** of pollution have changed, and now include more forms that are not readily broken down by the natural river processes we have discussed in earlier chapters. There are many forms of water pollution, ranging from bio-

logical waste products such as faecal material and sewage to manufactured toxicants used as herbicides or industrial cleaners. Each year, hundreds of new chemical compounds are brought onto international markets,
posing unforeseen pollution risks to aquatic ecosystems.

11.2.2 Sources and pathways of water pollution

The source of the pollutant is also relevant. Pollutants are often classified
according to whether they enter aquatic systems at **point sources** such as
pipes and localized entry sites or as **diffuse sources**, seeping in across a
broad interface (Fig. 11.1). Agricultural runoff from fields adjacent to
waterbodies exemplifies diffuse source pollution. The mode of entry of
pollution influences how easy it is to detect, control and manage. Point
sources such as effluent pipes can be more readily monitored and regulated than diffuse sources. Further, releases of pollutants at point sources
can be timed to coincide with periods of high discharge to dilute the
material. However, there is always the risk that some of this material will
settle out in the sediments along the river, constituting a diffuse source
later. For example, the World Heritage listed Kakadu National Park in
northern Australia sits on top of large, economically important mineral
reserves including uranium and gold. Mining for uranium occurs in the
catchments of seasonally flowing tributaries where a major environmental concern is the management of excess water that accumulates each year
within the mine sites due to monsoonal rain. These wastewaters contain
naturally occurring heavy metals and suspended solids at levels far

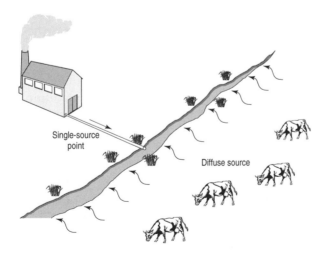

Fig. 11.1 A single-point source of pollution entering a river from a single pipe and a diffuse
source of pollution entering a river system from an area of catchment containing agricultural
development.

higher than background concentrations as well as synthetic chemicals such as hydrocarbons and process solvents. There is a detailed pollutant monitoring programme in the Park (Humphrey *et al.* 1990) but it is acknowledged that accumulation of metals in depositional areas such as the floodplain is a potential long-term hazard, despite carefully timed and regulated releases. Managers and ecologists must know the pathways of aquatic pollutants, both in the short term as well as over longer periods of decades and even centuries when sediment release occurs.

Pollutants are taken up by organisms through their food, via respiration, and by contact. Some pollutants have immediate effects whereas others accumulate slowly ('bioaccumulation') and their effects may not be evident for some time. Many synthetic pollutants such as organochlorines and polychlorinated biphenyls (PCBs) are **bioaccumulated**. As these compounds are long lasting, they are recycled through the food chain, increasing in concentration within animal tissues as they reach the upper trophic levels of top carnivores including humans (see below). The impacts of particular pollutants depend on their ecological pathways and whether or not they are immediately toxic. Their toxicity may also be increased during their metabolism by aquatic plants and animals.

11.2.3 Detecting water pollution: chemical vs. biological methods

How do we know what pollutants are in our waters? And how much? Most pollutants can be detected chemically, and in some cases, physically (as turbidity or water colour). This analytical chemical approach has been commonly used by water managers trying to monitor and regulate pollutant levels in standing and running waters. Aquatic pollution can also be detected by its effects on the biota in and around waterbodies, and the use of a suite of such ecological 'biomonitors' is growing increasingly popular (Harris & Silveira 1999). Large-scale biological monitoring of macroinvertebrates to assess river health is now conducted routinely in many countries worldwide (Metcalfe-Smith 1996), often by community groups and schools as well as management agencies. Other biota such as diatoms, fish and water plants also have potential as indicators of particular types of water pollution (Harris & Silveira 1999). The choice of the best indicator depends on many factors but is often a compromise of expertise, equipment, financial resources and existing data (Boulton 1999).

However, using biological indicators of pollution is not as simple as it sounds. As ecologists, we know that living organisms respond to many factors and it is often difficult to tease out the responses to a pollutant from all these other influences. One common approach has been to test the toxicity of particular pollutants in controlled experiments in laboratory tanks. Microcrustaceans such as water fleas (Cladocera) have been especially popular victims because they are small and easily maintained in laboratory cultures, they react rapidly (often within days) and there are standard protocols for their use (Chapman 1995). However, such data often only indicate short-term and threshold toxic effects, and extrapolation of results from the laboratory experiments to natural waterbodies is perilous because the pollutants seldom occur or act alone. Usually a polluted waterbody contains a cocktail of contaminants whose **cumulative** effects may be more toxic than their individual impacts. Plants and animals already stressed by low concentrations of toxicants may be especially prone to disease or death when exposed to mild pollution. Under these circumstances, predicting the threshold toxicity of pollutants based on laboratory findings is difficult and field experiments are needed to supplement these data.

11.3 PCBs in the Great Lakes

To illustrate some of the theory described above, let us explore the history of pollution of some of the Great Lakes of North America, focusing on how scientists have detected the effects of pollution in sportsfish and informed managers about ways to reduce the input and its effects. There are several key elements to this case study to keep in mind as you read through it. First, the pollutant we shall focus on, PCB, is a manufactured 'artificial' pollutant with a long period of persistence in the environment. The second is that like many examples of water pollution, the impacts were not evident until the pollution was quite severe. Third, the solution is likely to be complex, long term and involve a number of actions. Finally, the scientific approach to tackling the problem of pollution required research on the pathways and sources of the pollutant including its bioaccumulation up the food chain (Box 11.1). As so often is the case, humans and other vertebrates at the tops of these food chains are likely to receive the highest doses of pollutant, and some of the health effects of PCBs are tragic.

11.3.1 Typical microcosms of global aquatic pollution issues: the Great Lakes

Lake Erie, one of the Great Lakes, is some $25\,820\,km^2$ in area. In the 18th century, the small human population inhabited a virtually completely forested catchment interspersed with savannas of grass and oats. In many areas, huge marshes bordered the lake. By 1870, most of the forest had been cleared, grasslands were burned and some of the swamps drained. Erosion and siltation occurred, smothering fish spawning sites and aquatic vegetation. Meanwhile, human population in the catchment rose from about 3 million in 1900 to almost 15 million by 1980, and sewage was

Box 11.1

What is bioaccumulation?

Bioaccumulation of a pollutant occurs when we consume small amounts of a pollutant that we cannot break down or that we excrete only very slowly or not at all. Consequently, the pollutant tends to accumulate in our bodies over time. Over an extended period, the concentrations of the pollutant in our body can reach toxic levels impairing our health. Many synthetically produced chemical compounds have the potential to bioaccumulate, given that our bodies have had no previous exposure to them, and lack the means to either break them down or excrete them. Chemicals such as PCBs (see Section 11.3.1) are known to readily bioaccumulate in animal tissues. Organisms at the base of the food chain may absorb trace amounts of PCBs from the environment around them. Such small amounts are likely to be non-toxic and the organisms may be completely unharmed by the low concentrations of the pollutant in their body. A predator feeding upon this particular prey animal may, however, consume a huge number of individuals over its lifetime. Although the PCBs consumed when each prey animal is consumed is small, collectively the amount may be significant. If PCBs cannot be excreted, then the chemical will accumulate in the predator's body over time.

Bioaccumulation can be a significant problem for long-lived animals and for animals at the top of food chains. Clearly, the longer an animal lives, the longer the period of time over which the pollutant can accumulate. Hence, for very long-lived animals such as humans, the risk of bioaccumulating significant amounts of pollutant is high. For animals living at the top of a food chain, a process known as **biomagnification** can also significantly increase the amount of pollutant they consume and hence accumulate. An animal at the bottom of a food chain passes much or all of its accumulated pollutant on to the animal that consumes it. A predator that eats many prey effectively concentrates in its body the pollutant that its prey had collectively accumulated over their lifetime. At the next level up the food chain, a larger predator again consumes the pollutant previously concentrated in the first predator's body. This process occurs at each link in the food chain, with top predators ultimately consuming and biomagnifying the highest levels of pollutant in their bodies. Given that many top predators are also long-lived, the potential for the accumulation of toxic levels of persistent pollutants such as PCBs in their bodies is high. The combined processes of bioaccumulation and biomagnification of persistent pollutants have been implicated in declines in the abundance of a variety of top predators including several species of predatory bird and marine mammals.

routinely discharged into the lake. Between 1930 and 1965, average N and P concentrations increased three-fold. Biomass of phytoplankton in the lake trebled while rates of production increased by a factor of 20 (Regier & Hartman 1973). The bottom sediments of the lake turned anoxic as oxygen demand increased due to the settlement of detritus.

While Lake Erie was being impacted by the 'natural' pollutants of sediment and nutrients, a more insidious contamination by 'anthropogenic' or manufactured toxicants started around the middle of the last century. Many of these manufactured toxicants were byproducts of industrial processes making various type of plastics whereas others were herbicides or pesticides used in the Lake's catchment. Today, some 500 toxic compounds have been identified in the Great Lakes, entering via direct discharge into the water, stream runoff or atmospheric deposition (Laws 1993). The toxins of most concern are mercury and PCBs. PCBs are made by substituting up to 10 chlorine atoms onto a biphenyl ring structure (Fig. 11.2) with 209 different compounds possible. PCBs have

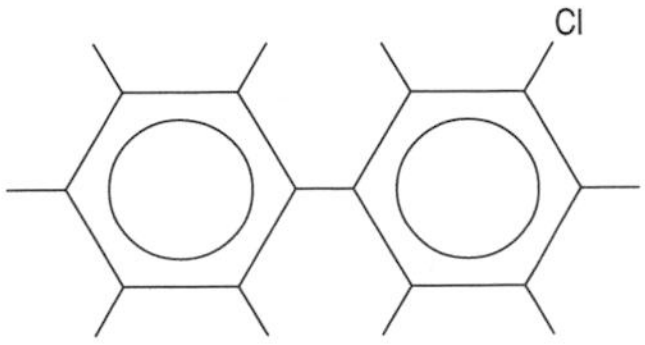

Fig. 11.2 A diagrammatic representation of a polychlorinated biphenyl (PCB) molecule. A chlorine (Cl) molecule can link to any of the 10 available positions, and one to 10 carbon atoms can be substituted onto the biphenyl aromatic structure to produce 209 different compounds collectively known as PCBs.

very high chemical, biological and thermal stability which make them very useful in industry – but they are also extremely persistent in the environment. They act like chlorinated pesticides in their toxicity to certain animals, interfering with the reproduction of many vertebrates including primates. In terms of human health, PCBs rank third in toxicity behind dioxins and furans and are the second most toxic pollutant in the Great Lakes (Bicknell 1992). Not until the early 1970s was this persistent toxicity realized, and in 1976 the Toxic Substances Control Act discontinued the production of PCBs in the USA with a complete ban in 1978 (Eby *et al.* 1997).

11.3.2 The tragic effects of the cocktail of contaminants

A theme in this chapter is the difficulty (and perhaps futility) of assigning 'blame' to a single pollutant, and indeed, often impacts occur due to the combination of toxicants. The first hints of the existence of pollutants that were targeting vertebrates, and possibly being concentrated along the food chain, became apparent as wildlife biologists began to document crashes of top predator fish populations in the Great Lakes as early as the 1940s. Lake trout declined sharply in the mid-1950s, and there were severe losses in colonial nesting waterbirds in the 1960s and 1970s (Colborn & Thayer 2000).

So, the first questions facing ecologists were to determine the causes of these declines and the sources of the pollutant(s). As the toxic properties of PCBs began to be better known, a comprehensive sampling programme commenced in the lake waters and sediments, and of fish and waterbird tissues. It quickly became clear that concentrations were locally high, and biomagnified up the food chain (see Box 11.1). Reaction and remedial legislation were swift and PCB usage in the catchment was banned in the mid-1970s.

Not surprisingly, there were several studies of human tissue levels of PCBs and other toxicants in the Great Lakes area. For ethical reasons, these studies have to be largely correlative (you cannot knowingly give humans toxic substances as part of an experiment!) so that direct experimental evidence for cause and effect links (see Chapter 1 on scientific method) cannot be obtained. Examination of various studies on the effects of eating fish with low levels of PCB contamination suggests the issue is far from resolved. Some studies indicate an elevated risk of developmental problems occurring in children exposed to PCBs. In contrast, other studies indicate a positive effect of having a diet rich in fish, even where the fish contain low levels of contaminants (see reviews by Stow *et al.* 1995b, Colburn & Thayer 2000).

11.3.3   The pathway of the pollutant

The next important question relates to the pathway of the pollutant. How do we limit its entry into the aquatic ecosystem, how do we minimize its effects on humans and aquatic life, and how long will the problem persist? Knowing a little about the chemistry of PCBs gives us a clue about their likely pathways. Many chlorinated organic molecules are far more soluble in fat than water and this is why PCBs accumulate so readily in fat (lipid) tissue. As breast milk is high in lipids, PCBs are readily transferred from mother to infant in this fashion. Similarly, fat stores in the eggs of fishes and birds are likely to contain excessive levels of PCBs because of their high solubility (Laws 1993).

Sources of PCBs in the food webs of the Great Lakes include atmospheric deposition (Sweet *et al.* 1993), resuspension from the lake sediments (Hermanson *et al.* 1991) and localized inputs from areas of landfill and industrial waste. While point sources from localized areas can be sealed with some success, there is currently no technically or economically feasible way to clean up atmospheric or sediment PCB sources to the Great Lakes (Jackson 1997). Meanwhile, PCBs travel through the food chain, via invertebrates, into fish. For most humans, the principal source of exposure to PCBs is consumption of contaminated fish. The evidence above demonstrates a 'concentrating effect' of PCBs as we move up the food chain so that top predators, including humans, are most at risk.

11.3.4   PCBs in fish – management of the Great Lakes fishery

Once the pathway was better understood, scientists could begin to examine the data on PCB levels in predatory fish and their prey to see whether concentrations were declining as a result of the methods used to restrict inputs of PCBs into the lakes. Fishing is a passion for many recreational users of the Great Lakes and these fisheries are intensively managed and stocked. To explore the trends in PCBs in fish in Lake Michigan, Stow *et al.* (1995a) analysed data on PCB concentrations in the tissue of fish caught from Lake Michigan since about 1974. For most fish species, the data indicated a decline from high levels in the mid-1970s to a levelling off in the mid-1980s (Fig. 11.3).

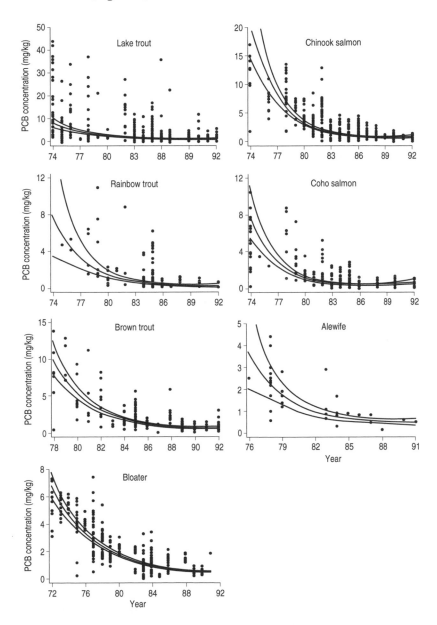

Fig. 11.3 PCB concentrations vs. year for each of seven fish species analysed in Lake Michigan. The centre line on each plot represents the best fit model (median value). Outer lines represent an interval of two standard error units on each side of the date. (After Stow *et al*. 1995 with permission of the Ecological Society of America.)

However, an interesting aspect of these data was the fact that two main prey fish species (alewife *Alosa pseudoharengus* and bloater *Coregonus hoyi*) varied substantially in their natural densities in the lake and hence, in their contribution to the diets of the top fish predators. As a result of this, Stow *et al.* (1995a) suggested that any predictive model of future trends in PCB concentrations must include these switches in feeding behaviours and changes in prey availability. It also provided a possible mechanism for managing the fishery so that the species that accumulate contaminants at the lowest levels could be stocked selectively.

However, like all scientific data, there were possible biases in the collection. Eby *et al.* (1997) considered the likely repercussions of changes in fish size and growth rates on the sampling programme being used to collect fish for tissue analyses. They discovered that lower growth rates of prey fish during the 1980s placed older, more contaminated prey fish in the size range most vulnerable to predators and in the size range sampled by the PCB monitoring programme. Eby *et al.* (1997) concluded that if trends in fish PCB concentrations were to be used as an indicator of trends in PCB pollution in the ecosystem, sampling must include a representative subset of fish of known age from the population. This sampling problem is important because unlike many of the other manipulative experiments described elsewhere in this book, predictive models at an ecosystem scale have been used as the main tool for informing managers on fish stocking rates. Obviously the models are only as good as the input data and our understanding of the key processes, and unrepresentative data jeopardize the accuracy of the predictions.

The sports fisheries in the Great Lakes are an important financial source for shoreline communities, and worth over $US100 million annually. The current fishery relies almost entirely on stocking but there have been some unexpected and rather unwelcome effects. In Lake Ontario, the combined effects of high rates of salmonid stocking with reduced inputs of phosphorus (controls of nutrient inputs) began to appear as signs of stress in the pelagic food web by 1990 (Jackson 1997). There were decreases in the size of salmon returning to spawn and reduced catches by anglers. Stocked salmon feed mainly on alewife and there were concerns that stocks of prey fishes could not meet demands by the sportfish. Salmonid predation had been so high that the alewife population comprised almost all pre-reproductive individuals (Jackson 1997). In response, salmonid stocking was reduced by around 55% between 1991 and 1994. However, the unexpected problem arises from the change in stocking policy. Reduced salmonid stocking to allow the alewife population to recover potentially leads to increased availability of larger and more contaminated prey, and hence, higher concentrations of PCBs in the salmonids. Ironically, the management of pelagic food webs in the Great Lakes to minimize the levels of PCBs reduces the sustainability of the lucrative sports fishery (Fig. 11.4). Jackson's modelling indicates that modest increases in stocking rates (ca. 25%) would decrease PCB concentrations in at least one salmonid species without a large increase in the probability that the prey alewife population would crash (Jackson 1997).

11.3.5 Persistence of PCBs in the Great Lakes

One of the major concerns about PCBs is their incredibly high persistence. This persistence would mean that serious environmental contamination with PCBs could not be reversed in the short term because the compounds would continue to be recycled in the ecosystem for many years. PCB production was banned in the USA in July 1979 (Laws 1993), and many point sources of PCBs in the Lakes were eliminated by the mid-1970s, resulting in decreases in PCB concentrations in many of the aquatic vertebrates (Scheider *et al.* 1998). Despite these initial declines, some fishes still have contaminant concentrations exceeding safe levels and in the 1980s, PCB concentrations seemed to level off rather than continuing to drop as was expected (Eby *et al.* 1997). Similarly, PCB concentrations in herring gull (*Larus argentatus*) eggs in the Great Lakes system have been logarithmically declining but rates of decline have stabilized or slowed at the Lake Superior colony, ceasing their decline in the mid-1980s (Pekarik & Weseloh 1998).

Most PCBs entering aquatic ecosystems are degraded in the water column, buried in sediments or volatilized (Laws 1993). It appears that in Lake Superior, volatilization rather than sedimentation is the

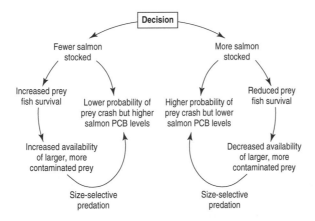

Fig. 11.4 Managing the Lake Ontario fishery for maximum sustainability and minimum PCB concentrations may represent a conflict of goals. Stocking fewer predatory salmonids decreases predation on prey fish, increases prey survival and leads to older more contaminated prey. Stocking more salmonids leads to young, less contaminated prey but insufficient prey are then available to support the larger salmon population. (After Jackson 1997 with permission of the Ecological Society of America.)

dominant pathway whereby PCBs are lost from the lake (Jeremiason *et al.* 1994). PCBs in the sediments may also be broken down by dechlorination by microbes in anaerobic environments (Beurskens & Stortelder 1995). Nonetheless, the losses are gradual and PCBs still pose a risk to humans even decades after their manufacture ceased. While good science underpins our understanding of the sources, pathways and ecological effects of these pollutants, social and political will is still needed to promote and fund this science, monitor the responses to management solutions and prevent recurrence of the problem from other pollutants.

11.4 Conclusion

Although management of the problems of pollution rest on good science, solutions have sometimes been more focused at treating the symptoms rather than the causes. For example, simply flushing the pollutant downstream (the early Thames example) is merely moving the problem elsewhere unless there are wetland processes downstream that are able to absorb and detoxify the wastes. In many cases, we have technological solutions that will reduce levels of pollution, especially from point sources, but they are expensive and frequently we need legislation to encourage polluters to comply. PCBs are just one example of the complex suite of pollutants that have been and, in many cases, continue to be intentionally or accidentally released into our waterways. What makes the PCB story sobering is that the 'lag effect' of this impact has persisted and sometimes becomes apparent in unexpected ways (e.g. via predator–prey shifts). As scientists applying our knowledge to understanding and managing aquatic systems, we need to be aware of these lag effects and indirect interactions – themes covered in earlier sections of this book.

11.5 Further reading

Bicknell D.J. (1992) Ranking Great Lake persistent toxics. *Water Environment and Technology* **4**, 50–55.

Beurskens J.E.M. & Stortelder P.B.M. (1995) Microbial transformation of PCBs in sediments – what can we learn to solve practical problems? *Water Science Technology* **31**, 99–107.

Boulton A.J. (1999) An overview of river health assessment: Philosophies, practice, problems and prognosis. *Freshwater Biology* **41**, 469–479.

Chapman J.C. (1995) The role of ecotoxicity testing in assessing water quality. *Australian Journal of Ecology* **20**, 20–27.

Colborn T. & Thayer K. (2000) Aquatic ecosystems: harbingers of endocrine disruption. *Ecological Applications* **10**, 949–957.

Collocott T.C. & Dobson A.B. (eds) (1974) *Chambers Science and Technology Dictionary*. Chambers, Edinburgh.

Eby L.A., Stow C.A., Hesselberg R.J. & Kitchell J.F. (1997) Modeling changes in growth and diet on polychlorinated biphenyl accumulation in *Coregonus hoyi*. *Ecological Applications* **7**, 981–990.

Harris J.H. & Silveira R. (1999) Large-scale assessment of river health using an Index of Biotic Integrity with low-diversity fish communities. *Freshwater Biology* **41**, 235–252.

Hermanson M.H., Christianson E.R., Buser D.J. & Chen L.M. (1991) Polychlorinated biphenyls in dated sediment cores from Green Bay and Lake Michigan. *Journal of Great Lakes Research* **16**, 121–129.

Humphrey C.L., Bishop K.A. & Brown V.M. (1990) Use of biological monitoring in the assessment of effects of mining wastes on aquatic ecosystems of the Alligator Rivers Region, tropical northern Australia. *Environmental Monitoring and Assessment* **14**, 139–181.

Jackson L.J. (1997) Piscivores, predation, and PCBs in Lake Ontario's pelagic food web. *Ecological Applications* **7**, 991–1001.

Jeremiason J.D., Hornbuckle K.C. & Eisenreich S.J. (1994) PCBs in Lake Superior, 1978–1992 – Decreases in water concentrations reflect loss by volatilization. *Environmental Science and Technology* **28**, 903–914.

Laws E.A. (1993) *Aquatic Pollution. An Introductory Text*, 2nd edn. John Wiley & Sons, New York.

Metcalfe-Smith J.L. (1996) Biological water-quality assessment of rivers: Use of macroinvertebrate communities. In: G. Petts & P. Calow (eds) *River Restoration*, pp. 17–43. Blackwell Science, Oxford.

Pekarik C. & Weseloh D.V. (1998) Organochlorine contaminants in herring gull eggs from the Great Lakes, 1974–1995: Change point regression analysis and short-term regression. *Environmental Monitoring and Assessment* **53**, 77–115.

Regier H.A. & Hartman W.L. (1973) Lake Erie's fish community: 150 years of cultural stress. *Science* **180**, 1248–1255.

Scheider W.A., Cox C., Hayton A., Hitchin G. & Vaillancourt A. (1998) Current status and temporal trends in concentrations of persistent toxic substances in sport fish and juvenile forage fish in the Canadian waters of the Great Lakes. *Environmental Monitoring and Assessment* **53**, 57–76.

Stow C.A., Carpenter S.R., Eby L.A., Amrhein J.F. & Hesselberg R.J. (1995a) Evidence that PCBs are approaching stable concentrations in Lake Michigan fishes. *Ecological Applications* **5**, 248–260.

Stow C.A., Carpenter S.R., Madenjian C.P., Eby L.A. & Jackson L.J. (1995b) Fisheries management to reduce contaminant consumption. *BioScience* **45**, 752–758.

Sweet C.W., Murphy T.J., Bannasch J.H., Kelsey C.A. & Hong J. (1993) Atmospheric depositions of PCBs into Green Bay. *Journal of Great Lakes Research* **19**, 109–128.

Chapter 12   *Can we fix smelly, green lakes?*

12.1   A lake turns green and smelly

Summer on the inland plains of south-eastern Australia is hot and sunny. It is perfect for growing a great variety of fruit, so long as there is access to plenty of water. Across the catchment of the Murray River system, many reservoirs have been constructed to store winter rains. Lake Mokoan is one of those lakes. When the dam that created Lake Mokoan was built, the stored water spread out over what was formerly rich grazing lands. The bleached skeletons of giant drowned river redgum trees still stand, rising from the still waters of the lake and reminding us of what was there before the area was flooded.

When the lake was first filled, dense beds of macrophytes grew over the former pasture, no doubt nourished by the abundant nutrients leaching from the rich sediment. The water was clear and could be used for a wide variety of purposes. However, over the summer of 1982–83 a severe drought struck. Winter rains had failed and over summer, Lake Mokoan was effectively emptied of water to supply the needs of irrigators. The macrophyte beds dried and turned to dust in the baking summer sun.

When the rains returned, Lake Mokoan refilled. The water now spread out over dusty claypans. When the wind blew, waves lifted the fine clay particles from the bed of the lake and into the water column. Introduced European carp had also found their way into the lake and further stirred up the lake sediments. The lake turned brown and little light penetrated to the bed of the lake. The nutrients in the water column triggered blooms of noxious planktonic blue–green algae. The water was now not only muddy, it smelt foul and was toxic to drink. Macrophytes that might have stabilized the lake sediments and drawn nutrients out of the water column were unable to establish due to the unstable substrates and lack of light reaching the lakebed. In a country that is often desperate for water and where the value of water can be counted in millions of dollars, the water stored in Lake Mokoan was now useless.

12.2 Using ecological theory to understand lake eutrophication

Algal blooms are primarily caused by an oversupply of nutrients. Algae requires nitrogen and phosphorus to grow. In many waterbodies, nutrients may be in short supply, particularly if the surrounding catchment is forested and nutrients are held tightly within the terrestrial ecosystem. However, in agricultural systems, farmers often spread fertilizer to increase crop productivity. Nutrients not taken up by crops wash into local waterways. Farm animals, particularly cows, may also deposit large amounts of nutrient-rich faeces and urine directly into the stream from which they drink. In urban environments, lawn clippings and pet faeces wash down stormwater drains providing another supply of nutrients. If those streams drain into a lake, nutrients can accumulate and turn it from an **oligotrophic state** (low nutrients, low productivity, clear water) or **mesotrophic state** (intermediate levels of nutrients, productivity and water clarity), to a **eutrophic state** (high nutrients, high primary production, low water clarity).

If nutrients are the key factor triggering algal blooms, then an obvious management strategy is to remove them. Nutrients entering a system from a point source can be controlled by removing the point source that is creating the problem, perhaps a sewage outfall. Unfortunately, nutrients frequently enter aquatic systems through non-point, diffuse inputs from across an entire catchment (see also Chapter 11). Control of such inputs can involve catchment management over potentially huge areas of land. Revegetation, education and the slow leaching of nutrients from soils that have been previously saturated with fertilizer can mean that control of non-point sources of nutrients can take decades. Any management approaches that have the potential to control algal blooms within a shorter time frame are very attractive.

12.3 Putting theory into practice: food webs, trophic cascades and biomanipulation

Does ecological theory suggest any alternative approaches to controlling nuisance algae? Understanding the factors that determine the abundance of organisms is, after all, a key goal of ecological science. Hairston *et al.* (1960) laid the theoretical groundwork that describes a phenomenon that has become known as a 'trophic cascade'. The existence of trophic cascades suggests that organisms linked by feeding or trophic relation-ships can be used to control each other.

Hairston *et al.* (1960) argued that an important determinant of an organism's abundance is its position within a food web and the configur-ation of the food chain within which it is embedded. In three-level food chains, that is primary producer–herbivore–predator, predators reduce the abundance of herbivores, thereby allowing the biomass of primary producers to increase. In such a situation, resource limitation is the primary constraint acting on the abundance of predators and primary producers. In contrast, where predators are removed and no control is exerted over herbivore populations, the biomass of primary producers at the base of the food chain will fall due to the impact of uncontrolled herbivore feeding. Hairston *et al.* (1960) observed that in terrestrial systems, outbreaks of herbivorous pests such as scale insects occurred when the insect predators of scale insects were removed. Similarly, herbivorous mammals such as deer reached levels of abundance that resulted in destruction of vegetation when predators were removed.

The arguments presented by Hairston *et al.* (1960) were largely directed at terrestrial food webs, however it is easy to extend their theoretical model to freshwater environments. Relatively simple food webs fre-quently exist in the pelagic zone of lakes (Fig. 12.1). Assuming sufficient nutrients are present, and suitable light and temperature conditions prevail, planktonic algae will form the base of the pelagic food web. This primary production supports filter-feeding, herbivorous zooplank-ton. Various species of zooplanktivorous fish and invertebrates then prey upon the zooplankton. Larger piscivorous fish, if present, feed upon the pelagic zooplanktivorous fish (see Chapter 4).

Application of the theoretical model of Hairston *et al.* (1960) to a pelagic food web leads to the prediction that the biomass of planktonic algae in a lake will be strongly influenced by the configuration of the pelagic food web (Fig. 12.1). In lakes supporting three trophic levels, algae will be abundant (Fig. 12.1a). Zooplanktivorous fish will reduce zooplankton, allowing the abundance of planktonic algae to increase. However, the addition of a fourth trophic level in the form of piscivorous fish, will reduce the abundance of zooplanktivorous fish, releasing zooplankton from predation (Fig. 12.1b). Increased zooplankton numbers result in reduced densities of planktonic algae. Theoretically, troublesome algal blooms could be controlled by manipulating the composition of the fish community, that is top–down control.

In an applied ecological context, the application of the trophic cascade model to the management of water quality in lakes has become known as **biomanipulation**. Broadly, biomanipulation can be defined as the re-structuring of the biological community to achieve reductions in algal biomass, improvements in water clarity and the promotion of diverse

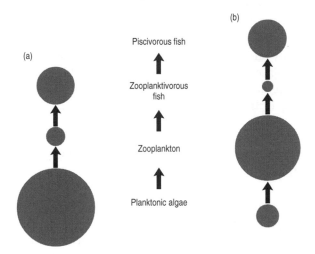

Fig. 12.1 Examples of pelagic food webs with three (a) and four (b) trophic levels. The size of the circles represents the biomass of each trophic level based on the theoretical predictions of Hairston *et al*. (1960). Arrows indicate the direction of the movement of materials through the food web.

biological communities (Perrow *et al*. 1997). In practical terms, it most frequently involves manipulation of fish communities with the aim of increasing zooplankton biomass and associated decreases in algal biomass (Perrow *et al*. 1997).

12.4 Do trophic cascades occur in lentic systems?

The concept of a trophic cascade is potentially attractive to those charged with managing water quality in lakes prone to algal blooms. It suggests that stocking a lake with piscivorous fish could result in reductions in the occurrence of algal blooms. In comparison, forcing changes to farming practices or sewering an entire neighbourhood with the ultimate aim of reducing diffuse nutrient inputs can require massive resources. Theoretically, top–down control of a food web may also result in relatively rapid control of algal blooms, given that efficient piscivorous fish may eliminate zooplanktivorous fish very quickly. Algal grazing zooplankton populations can also increase in numbers rapidly given suitable conditions.

Attractive theory must, however, be tested before it can be routinely recommended for use as a management tool. Testing whether trophic cascades occur presents a significant challenge given that it involves the manipulation of the biota of an entire lake. Several approaches have been used to test the occurrence of trophic cascades. Those approaches range from exploration of potential correlations between fish, zooplankton and algae across many different unmanipulated lakes, or from the one lake through time. Alternatively, many researchers have attempted to manipulate directly elements of lentic food chains, either within entire lakes, or smaller enclosures placed within a lake. The different approaches complement each other, with strengths and weaknesses being apparent in any of the possible research alternatives.

12.5   A classic correlative study: fish and zooplankton relationships in space and time

If zooplankton communities in lakes are structured by fish predation, then we might predict that zooplankton community structure will differ in a lake containing zooplanktivorous fish compared with a lake without zooplanktivorous fish. Further, we might also predict that the structure of zooplankton communities in fishless lakes will change following the introduction of fish. Those taxa vulnerable to fish predation will decline, whilst those that can successfully cope with fish predation will prosper. Determining the impact of fish predation can be done by comparing zooplankton communities in different lakes with and without fish. Alternatively, the zooplankton community within a single lake can be compared through time either before and after the introduction of zooplanktivorous fish, or vice versa.

In what is now considered to be a classic paper, Brooks and Dodson (1965) used both these approaches to compare zooplankton communities across lakes in southern New England. A number of lakes with and without alewife (*Alosa pseudoharengus*), a zooplanktivorous fish, were compared. The structure of the zooplankton community in a single lake before and after alewife were introduced was also examined. Marked differences between lakes with and without fish were evident. In lakes without alewife, large cladocerans and copepods such as *Daphnia* and *Mesocyclops*, respectively, tended to dominate the zooplankton community. In contrast, large *Daphnia* were absent from lakes with alewife. The zooplankton community in alewife lakes was dominated by small cladocerans, such as *Bosmina*, fast-moving copepods and rotifers. A similar pattern was repeated through time in Crystal Lake, with *Daphnia* and other large planktonic crustaceans disappearing from the lake following the introduction of glut or blueback herring (*Alosa aestivalius*), a species very similar to alewife (Fig. 12.2).

The most convincing aspect of the study by Brooks and Dodson (1965) is the repetition of the pattern of distribution of large slow-moving *Daphnia* in relation to lakes with and without fish, and before and after the introduction of fish. However, as with any descriptive study, and

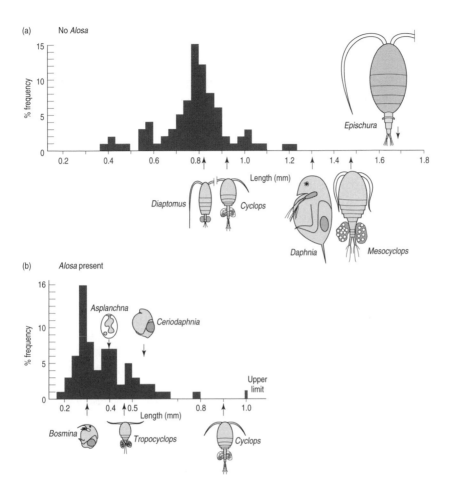

Fig. 12.2 The composition and size distribution of dominant zooplankton taxa in Crystal Lake before (1942) and after (1964) the introduction of alewife (*Alosa aestivalis*). Each bar of the histogram indicates that 1% of the total sample counted was within that size range. Arrows indicate the size of the smallest mature instar of each zooplankton taxon. (After Brooks & Dodson 1965 with permission from the American Association for the Advancement of Science.)

particularly one based on a relatively small number of samples, inference of a cause and effect relationship between any correlated variables must be viewed with caution. Factors other than those actually measured may have generated the observed patterns. The fact that the changes in the zooplankton communities observed by Brooks and Dodson (1965) correspond with our understanding of the fish–zooplankton dynamics, in that fish will tend to prey most heavily on the largest and slowest moving zooplankton taxon, that is *Daphnia*, lends weight to their argument. However, the possibility that other unknown factors may be determining

175

the abundance of both the alewife and the zooplankton cannot be absolutely discounted.

12.6 Meta-analysis: testing the trophic cascade model across systems

The analysis of Brooks and Dodson (1965) was restricted to a relatively small number of lakes across a limited area, and only examined planktivorous fish–zooplankton relationships. Their findings are interesting in that they suggest trophic cascades do occur, but they are limited for the purposes of biomanipulation. If biomanipulation does have potential as a management tool, trophic cascades must extend from piscivorous fish at the top of a food chain, all the way down to planktonic algae at the bottom. Further, if such cascades do occur, how common are they?

Since the original analysis of Brooks and Dodson (1965), several correlative analyses of fish–zooplankton–algal abundances across lakes and across different studies have been completed, enabling comparisons of patterns across geographical regions and types of lakes. Currie *et al.* (1999) combined the analysis of 29 small unmanipulated Canadian lakes with a review of five other across-lake comparisons of pelagic food webs, an analysis approach known as a **meta-analysis**. Pooling the results from several descriptive studies of lake food web structure can effectively increase the number of lakes sampled. Such comparisons require careful interpretation given that methods used across the different studies can vary greatly.

In their analysis of the 29 Canadian lakes they sampled, Currie *et al.* (1999) observed relationships that suggested that both top–down and bottom–up regulation of the pelagic food web was occurring. Chlorophyll a concentration, a measure of algal density, increased as nutrient (phosphorus) concentrations increased. The biomass and size of cladocerans also tended to increase in the presence of piscivorous fish, suggesting that predation by piscivorous fish was reducing planktivorous fish abundance, thus reducing the predation pressure on the cladocerans. However, they observed no relationships between the abundance of piscivorous fish and chlorophyll a concentration, suggesting that the effects of the removal or introduction of piscivorous fish were not cascading down food webs to the primary producers. Similar patterns were observed in the other multiple-lake analyses they reviewed, with relationships between adjacent trophic levels being observed, but no clear relationships existing between the abundance of organisms at either end of long food chains. The variability observed by Currie *et al.* (1999) in the structure of food webs across many lakes suggests that it is unrealistic to expect that simple manipulation of fish densities can be routinely used to control algae.

12.7 Experimental studies: controlled manipulation of pelagic food webs

The concept of trophic cascades lends itself to experimental testing. Removal or addition of one element of a food chain should result in predictable changes in the abundance of organisms within a food web. Such manipulations can potentially be conducted at the scale of the whole lake, or in enclosures within lakes. However, whole lakes are potentially difficult to manipulate. Controlling fish densities or nutrient inputs across a large lake can prove impossible. Replication of experimental lakes can also prove impossible, both in terms of obtaining sufficient lakes to manipulate, and due to the considerable variation that may occur in the physical and biotic structure of lakes, even within the same geographical region. Alternatively, mesocosms (experimental enclosures) can be used to contain fish and zooplankton or nutrients at experimental densities or concentrations. Precise control of experimental treatments can be maintained in small enclosures and replication of multiple experimental treatments is often possible. The potential cost of enclosures is the loss of realism.

12.7.1 Experimental mesocosms: pelagic dynamics in plastic bags

Lynch (1979) and Lynch and Shapiro (1981) were amongst the first to use small enclosures placed within a lake to test the question, does variation in fish density result in changes in zooplankton and phytoplankton communities? Eight polyethylene bags (1 m in diameter, 1.8 m deep and closed at the bottom) were suspended from the surface of Pleasant Pond, a 0.25-ha pond in Minnesota. The bags were filled with unfiltered lake water. Three of the enclosures were left fishless, and increasing numbers of bluegill sunfish (*Lepomis macrochirus*) were added to the remaining five enclosures (one, two, four or five fish – see Table 12.1). Over the next 6 weeks, the zooplankton and phytoplankton communities within each bag were sampled.

Table 12.1 Bluegill sunfish added and recovered from enclosures by Lynch and Shapiro (1981). Enclosures 1, 2 and 7 were fishless controls (not listed). With permission of the American Society of Limnology and Oceanography.

Enclosure	Added 18 June		Recovered 2 Aug	
	No. fish	Total wet weight	No. fish	Total wet weight
8	1	7.8	1	21.7
9	2	15.8	2	32.4
3	2	25.0	2	41.1
11	4	50.0	4	45.7
6	5	83.0	5	87.0

The presence or absence of fish in the enclosures clearly had a significant impact on the structure of the zooplankton and phytoplankton communities that developed within them (Lynch 1979, Lynch & Shapiro 1981). In bags without fish, large zooplankton, such as various species of *Daphnia*, tended to dominate. In contrast, in bags with fish, smaller zooplankton species such as *Bosmina* and *Cyclops* copepods were dominant (Fig. 12.3). The presence of fish in the experimental bags also had a marked effect on the algal community. Algal densities in the bags containing fish were between nine and 62 times higher compared with the bags in which no fish were present (Table 12.2). The highest

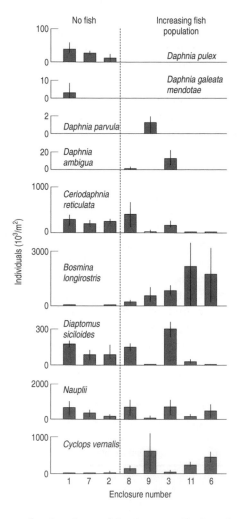

Fig. 12.3 Mean and range for abundance of dominant zooplankton for last three sampling dates of enclosure experiment. Refer to Table 12.1 for densities of fish per enclosure. (After Lynch 1979 with permission of the American Society of Limnology and Oceanography.)

Table 12.2 Total phytoplankton biomass in 10^5 μm^3/mL (mean for last three sampling dates +1 SD) in enclosures with varying fish densities. (After Lynch & Shapiro 1981 with permission of the American Society of Limnology and Oceanography.)

Enclosure	No. fish	Phytoplankton biomass
1	0	6.0+3.6
7	0	5.0+2.0
2	0	8.8+1.2
8	1	114.7+4.2
9	2	106.6+37.8
3	2	83.4+36.2
11	4	88.8+26.4
6	5	308.1+248.8

biomass of algae was recorded in the bag containing the most fish. Algal composition also changed markedly as the level of fish predation increased, with blue–green algae dominating at the highest fish densities.

The experiment reported by Lynch (1979) and Lynch and Shapiro (1981) suggests that trophic cascades can occur across at least three trophic levels, that is planktivorous fish–zooplankton–phytoplankton. However, their experiment had some obvious limitations, and their results cannot be directly extrapolated to the dynamics of entire lakes. Lynch and Shapiro used relatively small enclosures, and the likelihood of observing a strong fish effect in a small enclosure is high for at least two reasons. In small enclosures, the avoidance of predators by prey is difficult, hence predator–prey interactions tend to be artificially intensified. For that reason alone, it is unrealistic to conduct long-term studies of interactions between piscivorous and planktivorous fish in small enclosures. The big fish will almost inevitably eat all the small fish, something that does not always happen in real lakes. Further, fish also introduce nutrients into the water column as they excrete wastes. Their contribution to nutrient loading can be significant in a small experimental enclosure. Enclosure experiments can indicate whether trophic cascades might occur, and the mechanisms that might drive such changes. However, such experiments cannot tell us whether similar changes will occur in real lakes. Only an experiment conducted in a real lake can do that.

12.7.2 Manipulating whole lakes

Carpenter *et al.* (1987) describe a particularly ambitious whole-lake experiment in which they either manipulated or compared pelagic food webs across three morphometrically similar lakes (Paul, Peter and Tuesday Lakes) lying within 1 km of each other. Each lake was surrounded by similar protected catchments to minimize the wider effects of catchment variation. Paul Lake was maintained as an unmanipulated control, enabling the effects of interannual climatic variation on undisturbed lake communities to be observed. Approximately 90% of the adult largemouth bass population (*Micropterus salmoides*) in Peter Lake, was transferred to Tuesday Lake, and approximately 90% of the minnow (*Phoxinus eos, P. neogaeus* and *Umbra limi*) biomass in Tuesday Lake was transferred to Peter Lake. These transfers were intended to produce low piscivore/high planktivore and high piscivore/low planktivore treatments in Peter and Tuesday Lakes, respectively (Table 12.3).

The logical strength of the multi-lake manipulation experiment completed by Carpenter *et al.* (1987) is immediately apparent when the pelagic communities present in unmanipulated Paul Lake from 1984 and 1985 are compared. The summer of 1985 was cooler than the summer of 1984 resulting in reduced primary productivity in 1985. This resulted in lower algal biovolume and zooplankton biomass in 1985 compared with 1984 (Figs 12.4 & 12.5). Such a result is important, as it provides a clear indication that substantial interannual variation in pelagic communities can occur, irrespective of any experimental manipulations. Such natural interannual variation can be compared to the effects of the experimental treatments, enabling the additional effect of the experimental treatment on lake community structure to be identified.

In Tuesday Lake (high piscivore/low planktivore treatment), the combination of bass stocking and minnow removal was expected to reduce the zooplanktivory, resulting in a decrease in phytoplankton densities. The manipulation of the fish occurred as intended, resulting in the effective elimination of minnow populations. The effect of this change in the fish community on the wider pelagic community corresponded with the predictions of the trophic cascade model. The total zooplankton biomass increased markedly in August 1985 (Fig. 12.4). In contrast, algal biovolumes were markedly lower in 1985 compared with 1984 (Fig. 12.5). This result could, however, be a result of the interannual variation in climate between 1984 and 1985. However, plotting the differences

Table 12.3 Mean fish densities (SE in parentheses) in the study lakes in August 1984 and August 1985. (After Carpenter *et al.* 1987 with permission of the Ecological Society of America.)

	Paul		Peter		Tuesday	
	1984	1985	1984	1985	1984	1985
Minnows (catch/trap/h)	0	0	0	0	0.94(0.22)	0
Bass (fish/ha) YOY	−*	242 (225)	−*	5960 (1120)	0	4200 (2020)
Juveniles (< 195 mm)	80 (10)	187 (48)	67 (5)	19 (10)	0	190 (80)
Adults (> 195 mm)	201 (87)	162 (54)	154 (61)	16 (3)	0	293 (91)

*In 1984, young of the year (YOY) were present in Paul and Peter Lakes, but catches were too low to estimate densities.

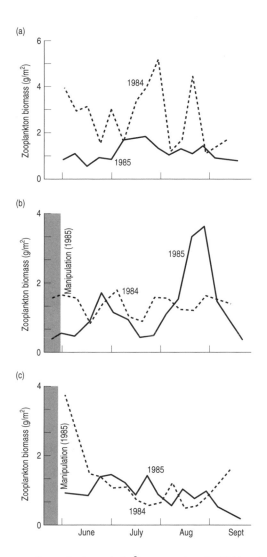

Fig. 12.4 Dry biomass of zooplankton (g/m^2) versus date in 1984 and 1985 in (a) Paul, (b) Tuesday and (c) Peter Lake. The shaded bar indicates the period of manipulation. (After Carpenter *et al.* 1987 with permission of the Ecological Society of America.)

between the pelagic communities in Paul and Tuesday Lakes in 1984 and 1985 revealed that zooplankton densities increased markedly in Tuesday Lake relative to Paul Lake after the addition of piscivores and removal of planktivores (Fig. 12.6). Whilst primary productivity declined in both lakes in late summer, the decline was relatively greater in Tuesday Lake in the late summer of 1985 when the intensity of grazing was at its highest. This pattern suggests an influence on pelagic communities in Tuesday Lake that was additional to the changes related to climate.

181

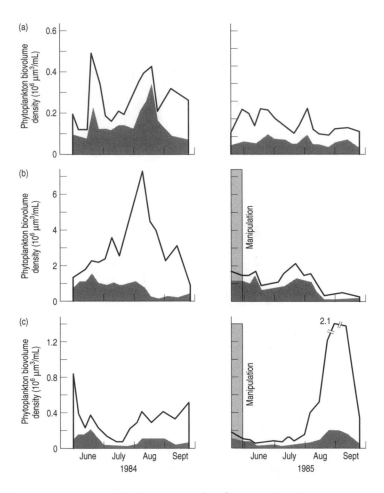

Fig. 12.5 Phytoplankton biovolume density (10^6 μm^3/mL) versus date in 1984 and 1985 in the three study lakes: (a) Paul, (b) Tuesday and (c) Peter Lake. The biovolume is divided into small (shaded) and large (open) size classes. (After Carpenter *et al.* 1987 with permission of the Ecological Society of America.)

In Peter Lake (low piscivore/high planktivore), the combination of low bass and high minnow numbers was expected to depress zooplankton densities resulting in high densities of phytoplankton. However, whilst phytoplankton did increase as predicted (Fig. 12.5), the mechanism leading to that response was not as expected. The large numbers of minnows introduced into Peter Lake were largely eliminated from the lake by the largemouth bass that remained after the attempted removal (10% of biomass remained after removal attempts). Consequently, after late June, few minnows were observed in the lake. In the absence of the planktivorous minnows, zooplankton densities might have been expected to increase, resulting in a drop in algal abundance. However, largemouth

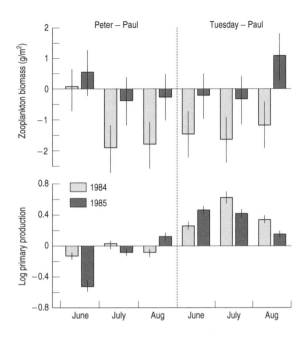

Fig. 12.6 Interlake differences between manipulated and reference ecosystems in zooplankton biomass and primary production in 1984 and 1985. The vertical line through each bar represents the 95% confidence interval for the mean difference between lakes. (After Carpenter *et al*. 1987 with permission of the Ecological Society of America.)

bass spawned in Peter Lake as the piscivore removals were occurring. In the absence of large numbers of cannibalistic adult fish, survival of the young of the year (YOY) fish was very high. It was these YOY fish that preyed primarily upon zooplankton and resulted in low zooplankton abundances (Fig. 12.4). In contrast to the low algal densities observed in the other two lakes, algal densities increased markedly in late summer in Peter Lake (Fig. 12.5). The study demonstrated that manipulation of fish populations can cascade down to lower trophic levels in whole lakes. However, the mechanisms involved could never have been predicted from studies conducted within the limited confines of a small enclosure.

12.8   Reviewing the results of fish manipulation experiments

The experiment described by Lynch (1979) and Lynch and Shapiro (1981) represents one of the earliest attempts at manipulating pelagic food webs within enclosures. Since their experiment, numerous enclosure experiments have been completed. A number of other whole-lake experiments have also been conducted in addition to the study conducted by Carpenter *et al.* (1987), although it is significant that few multi-lake experimental studies have been conducted. Most whole-lake studies have examined temporal change in pelagic communities following manipulation of either fish or nutrients in a single lake, a reflection of the difficulty of gaining access to more than one lake, or the logistical problems associated with manipulating more than one lake.

The number of published enclosure and whole-lake experimental studies has enabled multi-study comparisons, similar to those comparing pelagic communities in unmanipulated lakes. DeMelo *et al.* (1992) reviewed the results of 18 enclosure and 26 whole-lake experiments. They specifically examined the size range of enclosures and lakes used for study, the duration of experiments, the strength of the experimental manipulations and the trophic state of the lakes studied (oligotrophic vs. eutrophic lakes). The strength of the top–down cascade in each experiment was scaled from zero to four. A score of zero indicated that planktivore–zooplankton–phytoplankton responses to the experimental manipulation were totally contrary to the predictions of the trophic cascade model. A perfect score of four indicated perfect agreement with predictions.

A key pattern that emerged from the meta-analyses conducted by DeMelo *et al.* (1992) was the high degree of variation in the strength of top–down responses that occurred following experimental manipulations of food webs. No obvious patterns were apparent in the data, other than a slight tendency for weaker top–down responses to occur following experimental manipulations in large lakes and over long periods (Fig. 12.7). DeMelo *et al.* (1992) concluded rather pessimistically that biomanipulation could not be recommended as a reliable water quality management tool given the uncertainty regarding the outcome of any particular manipulation. Numerous confounding influences, such as excretion of nutrients, presence of prey refuges, effects of bioturbation by fish, sediment resuspension and the edibility of algae, complicated the responses of pelagic food webs to experimental manipulations.

12.9 Does biomanipulation work?

The review by DeMelo *et al.* (1992) generated considerable debate in the literature regarding the occurrence of trophic cascades in pelagic systems and the value of biomanipulation as a management tool. Carpenter and Kitchell (1992) argued that too few well-designed, large-scale, whole-lake manipulations had been conducted to assess fully the potential of biomanipulation. They pointed out that strong trophic cascades do occur, as observed by DeMelo *et al.* (1992), however the lack of rigorous studies across a range of lake types meant that the conditions under which strong cascades occur remains unclear.

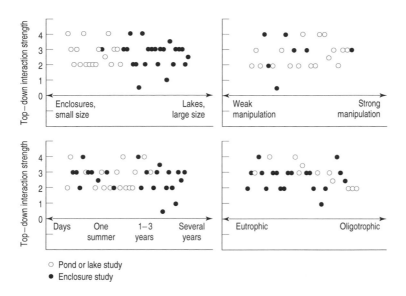

○ Pond or lake study
● Enclosure study

Fig. 12.7 Correspondence between theoretical predictions and real observations following experimental manipulations of pelagic food webs in lakes. A score of four indicates response is in perfect agreement with top–down/biomanipulation theory. A score of zero indicates complete disagreement. (After DeMelo *et al.* 1992 with permission of the American Society of Limnology and Oceanography.)

A number of whole-lake manipulations have been published since the review by DeMelo *et al.* (1992). The results of these studies tend to confirm the emerging trend, that is that manipulation of the fish communities at the top of pelagic food webs can have a variety of often complex impacts on the structure of pelagic communities. Removal of planktivorous fish has been shown to lead to increases in zooplankton and decreases in algal density. However, in several studies this effect has been difficult to maintain due to fish recolonization, or compensatory increases in the abundance of other zooplanktivores (Meijer *et al.* 1994, Elser *et al.* 1995, Moss *et al.* 1996). Horpila *et al.* (1998) removed planktivorous fish, resulting in a decline in nutrients and algal concentrations. However, zooplankton densities were not affected, suggesting that the effect of fish in their study lake on nutrient dynamics was significant. Persson *et al.* (1993) manipulated piscivore and planktivore fish numbers in a highly productive lake. The phytoplankton community responded only weakly to the manipulations of fish densities suggesting that the effects of the fish manipulation weakened at lower trophic levels.

The pelagic environment of lakes was once viewed as a relatively simple, tightly linked system consisting of piscivores–planktivores– zooplankton– phytoplankton. That simplistic view of pelagic systems is increasingly being revised as an understanding of the various interactions that occur among different species and the wider environment are described. Detailed reviews by Reynolds (1994) and Perrow *et al.* (1997) of where biomanipulation might be used to improve water quality, and of the major gaps in our current state of knowledge have emphasized that complexity. The effect of sediment resuspension in shallow lakes has emerged as a key factor that may limit the use of fish as a biomanipulation tool. Where sediment is regularly resuspended by wind and wave action, manipulation of piscivorous fish communities may have little impact on water quality (Scheffer *et al.* 1993). The use of biomanipulation as a management tool needs to be approached on a case-by-case basis, rather than being seen as a simple cure for water quality problems in eutrophic lakes.

12.10   Further reading

Brooks J.L. & Dodson S.I. (1965) Predation, body size, and composition of plankton. *Science* **150**, 28–35.

Carpenter S.R. & Kitchell J.F. (1992) Trophic cascade and biomanipulation: interface of research and management – a reply to the comment by DeMelo *et al. Limnology and Oceanography* **37**, 208–213.

Carpenter S.R., Kitchell J.F., Hodgson J.R. *et al.* (1987) Regulation of lake primary productivity by food web structure. *Ecology* **68**, 1863–1876.

Currie D.J., Dilworth-Christie P. & Chapleau F. (1999) Assessing the strength of top-down influences on plankton abundance in unmanipulated lakes. *Canadian Journal of Fisheries and Aquatic Sciences* **56**, 427–436.

DeMelo R., France R. & McQueen D.J. (1992) Biomanipulation: hit or myth? *Limnology and Oceanography* **37**, 192–207.

Elser J.J., Luecke C., Brett M.T. & Goldman C.R. (1995) Effects of food web compensation after manipulation of rainbow trout in an oligotrophic lake. *Ecology* **76**, 52–69.

Hairston N.G., Smith F.E. & Slobodkin L.B. (1960) Community structure, population control and competition. *American Naturalist* **94**, 421–425.

Horpila J., Peltonen H., Malinen T., Luokkanen E. & Kairesalo T. (1998) Top–down or bottom–up effects by fish: issues of concern in biomanipulation of lakes. *Restoration Ecology* **6**, 20–28.

Lynch M. (1979) Predation, competition and zooplankton community structure: an experimental study. *Limnology and Oceanography* **24**, 253–272.

Lynch M. & Shapiro J. (1981) Predation, enrichment, and phytoplankton structure. *Limnology and Oceanography* **26**, 86–102.

Meijer M-L., Jeppesen E., van Donk E. *et al.* (1994) Long-term responses to fish stock reduction in small shallow lakes: interpretation of five-year results of four biomanipulation cases in The Netherlands and Denmark. *Hydrobiologia* **275/276**, 457–466.

Moss B., Stansfield J., Irvine K., Perrow M. & Phillips G. (1996) Progressive restoration of a shallow lake: a 12-year experiment in isolation, sediment removal and biomanipulation. *Journal of Applied Ecology* **33**, 71–86.

Perrow M.R., Meijer M.-L., Dawidowicz P. & Coops H. (1997) Biomanipulation in shallow lakes: state of the art. *Hydrobiologia* **342/343**, 355–365.

Persson L., Johansson L., Andersson G., Diehl S. & Hamrin S.F. (1993) Density dependent interactions in lake ecosystems: whole lake perturbation experiments. *Oikos* **66**, 193–208.

Reynolds C.S. (1994) The ecological basis for the successful biomanipulation of aquatic communities. *Archiv für Hydrobiologie* **130**, 1–33.

Scheffer M., Hosper S.H., Meijer M.L., Moss B. & Jeppesen E. (1993) Alternative equilibria in shallow lakes. *Trends in Ecology and Evolution* **8**, 275–279.

Chapter 13   *What is the impact of introduced species?*

13.1   Finding the same fish wherever we go!

One of the most exciting aspects of studying ecology is simply appreciating biodiversity. To visit a new location, and to explore a range of species that you have never seen before is something special. Several years ago I (G.C.) visited Sumatra, one of the main islands of Indonesia. Sumatra sits on the Equator, and as is the case with many tropical regions, is a major centre of biodiversity. A total of 232 species of freshwater fish has been recorded in Sumatra, many of which are endemic to the region. The visit was an exciting opportunity to see a wide variety of fish that I had never seen before.

The first opportunity I had to collect some wild fish was at Lake Toba, a massive volcanic lake south of the city of Medan. We arrived on the shores of the lake at night. Next morning, the sight of a deep, blue lake surrounded by steep volcanic slopes covered in rough pasture and remnant tropical forest greeted me. Walking along the shore of the lake, I could see small fish darting across the rocky substrate ahead. Grabbing some wire and nylon netting from the nearby hotel, I quickly made a small dip net, and then sat waiting on a rock for some fish to swim by.

The fish were cautious and fast moving, and not easy to catch. After 20 min, I finally captured one small individual in the net. I carefully unwrapped my prize. The fish in the bottom of the net was totally familiar. I had last caught the same species only a week before in Australia. Three years later I would observe and catch the same species on the North Island of New Zealand. It was a mosquitofish or *Gambusia*, a species that has been widely spread around the world as a potential mosquito control agent.

Catching the same species of fish wherever I go seems to be something of a pattern. I grew up in England catching perch, roach and trout. I moved to Australia, and caught perch, roach and trout (along with

Gambusia), all of which had been introduced many years before. I have since moved to New Zealand, where I can also catch perch, trout and *Gambusia*, again all introduced over the past 150 years. Slowly, we are losing the excitement of visiting a new location and having the opportunity to explore a totally new fauna. With each new introduced species, the world becomes a little less exciting. With each new introduction, somewhere different becomes just a little more like everywhere else.

13.2 What are the effects of introduced species?

The effect of introduced species on ecosystems provides an excellent opportunity to integrate much of the material covered in previous chapters. The precise impact will depend on many factors. These might include the capacity of invading species to disperse into new habitats, physiological tolerance, fecundity, the range and type of resources required, capacity to monopolize resources and likely impact on habitat structure. Introduced species may reduce the abundance of preferred prey species (see also Chapter 9) or other species with which they compete. Alteration of the physical and chemical conditions within a habitat may have a variety of indirect effects on other species requiring a particular set of conditions. Not all species will decline following the invasion of a new species. Some may benefit, particularly if they can prey upon the new invader or perhaps tolerate the new environmental conditions created.

Gambusia can be considered to be a highly successful invader. They originated in south-eastern North America, but are now widespread through much of south-east Asia, many Pacific Islands, southern Europe, Australia and northern New Zealand. They possess many of the characteristics that make a particular species a successful invader. They can tolerate a wide range of environmental conditions. Their reproductive capacity is high (fish can mature in under 2 months), and they have an impressive ability to disperse widely. They tend to monopolize resources that they require, often at the expense of other less competitive species. Their ability to tolerate high temperature, low oxygen conditions that would kill many other fish, along with their capacity for rapid reproduction suggests that they are a potentially useful mosquito control agent. As a result they have been introduced into a wide range of temporary wetlands and other lentic habitats in tropical and subtropical areas, often displacing a local fish fauna that may have provided perfectly adequate or even superior mosquito control.

Once present in an region, *Gambusia* tend to spread so quickly and widely into all habitats that detailed study of their impact is difficult. For

example, locating lowland wetlands free of *Gambusia* across the vast
Murray River floodplain in south-east Australia is virtually impossible.
Detailed before and after studies of their impact on the Murray River
wetlands were never completed. Consequently, their precise impact can
only be surmised using anecdotal accounts of events around the time of
the invasion (and that date is uncertain), and by comparing conditions in
the very limited number of *Gambusia*-free wetlands with conditions in the
majority of wetlands that now contain *Gambusia*. If we truly want to
understand the full range of effects that an introduced species can have,
then we must complete detailed pre- and post-introduction studies that
measure the factors that the new species is likely to alter. Given that most
introductions of new species are unexpected, being either accidental or
illegal, few pre- and post-invasion studies are ever conducted.

13.3   The invasion of the zebra mussel

Zebra mussels (*Dreissena polymorpha*) are small (30–40 mm adult shell
length), filter-feeding bivalve mussels. They attach to hard substrates in
lakes and rivers by byssal threads. Zebra mussels are enormously fecund,
with the production of eggs by adult females ranging between 30 000 and
1.5 million eggs. The eggs hatch into a planktonic veliger larvae which
can drift on water currents for up to 5 weeks (Fig. 13.1). The combination
of high fecundity and a planktonic phase in their lifecycle provides them
with the potential to spread long distances downstream within river
systems. Depending on temperature and food availability, breeding
may occur in the first year of life, and individual mussels can live for
up to 9 years.

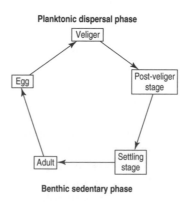

Fig. 13.1 The lifecycle and developmental stages of the zebra mussel (*Dreissena polymorpha*).
(After Mackie 1991 with kind permission of Kluwer Academic Publishers.)

Prior to 1800, zebra mussels had a distribution that was largely restricted to the Black, Caspian and Azov Seas in eastern Europe. In the 170 years that followed, they expanded their range to encompass most of Europe. Larval veligers disperse long distances by downstream drift, however natural transfers between river basins occur rarely due to the inability of zebra mussels to tolerate salinities greater than four parts per thousand.

The spread of zebra mussels across Europe from the 1800s onward coincides with a massive increase in speed and extent at which humans began to move themselves and materials between river basins across Europe. Transport canals linking river basins were built, enabling not only interbasin transfer of boats but also water and planktonic organisms. Water may be transported in the hulls of ships and boats as bilge or ballast water. Survival and the spread of organisms through the dispersal of this water increased significantly as travel times between ports fell with the advent of the steam engine. Mussels can also survive for days in wet fishing equipment or anchor ropes, invading new habitats when the equipment is used again. As travel times decreased, the likelihood of survival and transfer of mussels from one location to another increased.

13.3.1 Zebra mussels invade North America

On 1 June 1988, Sonya Gutschi and Ron Allison collected a zebra mussel whilst sampling in Lake St Clair, a small lake in the Great Lake system of North America lying between Lakes Huron and Erie (Fig. 13.2). Zebra mussels had somehow crossed the Atlantic Ocean. The presumed mechanism of transfer was via the release of veliger larvae in the ballast water of ships sailing from Europe into the Great Lakes system via the St Lawrence Seaway. The risk of such transfers occurring had been highlighted several years previously in an Environment Protection Service (1981) report. The discovery of zebra mussels immediately raised a number of questions that will confront any ecologist facing the potential invasion of a new species. The questions include:

1 How widespread is the newly discovered introduced species and have they started to reproduce?
2 How far will they spread?
3 What are the likely mechanisms of dispersal?
4 What are their likely ecological impacts?

13.3.2 How widespread are zebra mussels, and are they reproducing?

The discovery of a small number of individuals of an exotic species does not mean that the species will invade. Introductions of exotic species often fail for a variety of reasons, including unsuitable environmental

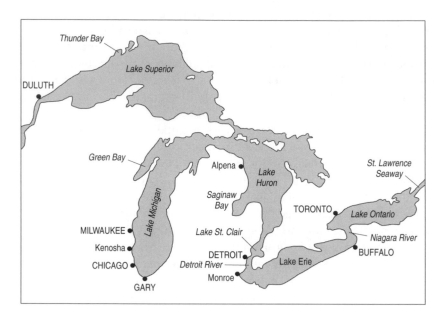

Fig. 13.2 Map of the Great Lakes.

conditions, insufficient resources or simply too few initial colonists. The collection of zebra mussels in 1988 presented ecologists with several scenarios. The mussels may have represented the survivors of a small population that were unable to breed and would subsequently die out. Alternatively, the population may have been a very recent addition to the Great Lakes fauna that had not yet bred nor spread beyond the point of initial introduction. If so, it might have been possible to contemplate some form of control or even elimination. Finally, the zebra mussels that were collected may have been a part of a much larger population that had been breeding successfully for some years, had dispersed away from the point of initial colonization and were now an established part of the lake fauna. Any control attempts would be pointless under this last scenario.

In early August 1988, an initial study to determine the extent of the zebra mussel invasion of Lake St Clair was commenced (Hebert *et al.* 1989). Key aims of the study were to determine the distribution of the mussel across the lake, and whether the population had successfully bred. Determining the distribution of the mussel was addressed by selecting and surveying 10 sites across the lake. Selection of the sites was stratified to ensure that the dominant substrates within the lake were sampled. Time since initial colonization and the frequency of successful breeding events was examined by measurement of the shell length of all mussels collected at the

10 sites with the aim of examining population structure and determining whether juvenile mussels were present.

The results of this initial survey demonstrated that zebra mussels were indeed well established in Lake St Clair (Hebert *et al.* 1989). The species was recorded from four of the 10 sites surveyed (Fig. 13.3a). Their distribution was restricted to the south-east region of the lake, however examination of their population structure suggested that at least one successful breeding event had occurred (Fig. 13.3b). The population had

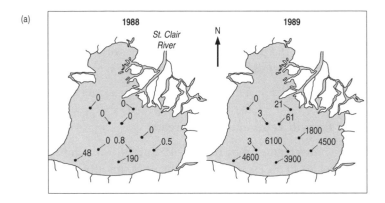

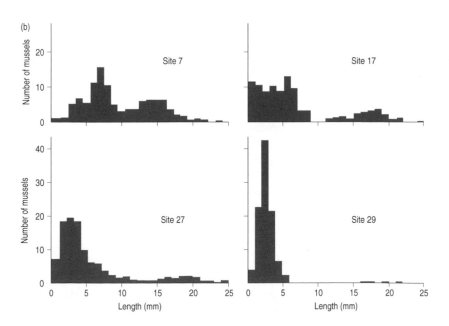

Fig. 13.3 A comparison of (a) the density (individuals/m^2) of zebra mussels at 10 sites in Lake St Clair sampled in 1988 and 1989, and (b) the size distribution of zebra mussels at four sites in Lake St Clair in August 1989. (After Hebert *et al.* 1991 with permission of the National Research Council of Canada.)

a bimodal distribution with mussels under 7.7 mm in length being most abundant. No individuals between 7.7 and 18.0 mm in length were collected, although a total of 25 individuals over 18.0 mm in length were collected. Individuals larger than 20 mm were sexually mature, and the largest individuals in the population were approximately 3 years old as determined by examination of shell sections.

The results suggested that zebra mussels had first colonized the lake in 1985, a pattern consistent with the results of previous surveys of the lake in 1983 during which no zebra mussels were recorded. The population had initially failed to reproduce successfully, however an abundance of juveniles in the population suggested that breeding in early 1988 had been successful, and the potential for the further spread of zebra mussels existed. Given the area across which the mussels were distributed, control was not an option.

13.3.3 How far will zebra mussels spread?

Understanding the factors that might ultimately limit the spread of an introduced species is an essential element in developing a management programme. Knowledge of the resource requirements of the invading species, and the physical and chemical extremes it can tolerate, permits identification of the waters that are potentially at risk of invasion.

Strayer (1991) conducted the first detailed study aimed at identifying North American freshwater habitats potentially at risk of zebra mussel invasion. Strayer (1991) conducted a meta-analysis of data that had previously been published on the distribution of zebra mussels across Europe where the species was long established and widely distributed. Strayer (1991) reasoned that the distribution of zebra mussels in Europe could provide crucial information on the ultimate distribution of the mussel in North America given the similar latitudinal spread, climatic and other environmental conditions across the two continents. Analysis of data from Europe would allow description of the potential range of zebra mussels at two levels of scale (see Chapter 2): first, the species' ultimate geographical spread as determined by extreme environmental conditions, and second the factors determining the species distribution within that range.

To determine the climatic factors that might limit the distribution of zebra mussels, Strayer (1991) plotted records of zebra mussel across Europe using the information derived from 44 studies. The distribution of mussels was then compared with a variety of climatic statistics derived from 110 weather stations across Europe. Weather stations were considered to lie within the range of zebra mussels if they were located within 100 km of the nearest zebra mussel record. Strayer (1991) concluded that zebra mussels were distributed across most of Europe, suggesting that the mussels could tolerate the range of climatic conditions found across Europe. The absence of zebra mussels from the Iberian Peninsula, Scandinavia and Italy was either due to a failure to disperse into these areas, or factors other than climate given that the mussels tolerated comparable climatic conditions elsewhere in Europe. Based on the range of conditions tolerated by zebra mussels in Europe, Strayer (1991) concluded that zebra mussels could potentially colonize most of North America.

Strayer (1991) also examined the occurrence of zebra mussels in lakes across Europe for which data on water quality existed. Zebra mussels were most often found in relatively large, deep, hardwater (high dissolved calcium), clear lakes. However, such conclusions must be viewed as tentative given that they were based on a small data set: 24 lakes in which zebra mussels were present and six lakes from which they were absent. Zebra mussels tended to be less abundant or absent in small shallow lakes where either winterkills or predation by birds may have limited their abundance. **Water hardness** is a measure of the calcium content, and it is well known that the distribution of many molluscs is limited by their high requirement for calcium, which is essential for shell building. Strayer (1991) observed that most European waters are relatively hard (>1 mmol Ca^{2+}/L), with the notable exception of Scandinavia where soft waters predominate and zebra mussels were absent. Strayer (1991) suggested that low dissolved calcium levels in many waters across North America might limit the abundance of the invader.

13.3.4 How do zebra mussels spread?

Whether a species will invade a suitable waterbody will depend on whether the species can disperse into that habitat. Understanding the mechanisms involved in the dispersal (see Chapter 6) of an invading species is another key step in developing control strategies. Indeed, the rate of spread may be slowed or even stopped if management strategies that prevent the transfer of the introduced species from one waterbody to another can be developed. The experience of almost 200 years of range expansion in Europe demonstrated that zebra mussels are spread by the downstream drift of planktonic larvae, and through the deliberate or accidental transport of adults or larvae by humans. Based on the European experience, it could be predicted that the spread of zebra mussels in North America would be characterized by the appearance of an isolated colonizing population, followed by a downstream expansion of range. Knowing the initial limited distribution of the mussel in 1989 allowed ecologists to maintain detailed maps of zebra mussel distribution over the following years.

Griffiths *et al.* (1991) plotted all records of zebra mussel sightings made during periodic surveys of likely mussel habitats throughout the Great Lakes between 1989 and 1990. As predicted, the distribution of the mussels expanded downstream from the initial point of introduction, spreading out from Lake St Clair, across Lake Erie, and then into Lake Ontario by 1989 (Fig. 13.4). From 1989 onwards, isolated populations located upstream of the initial introduction appeared, first in Lake Huron, and then in Lake Michigan. Dispersal associated with human activities such as commercial shipping, boating and fishing probably accounts for occurrence of these isolated populations. Once present in Lake Michigan, it was expected that the mussel would spread downstream into the Mississippi River via the Chicago Sanitary and Ship Canal and Illinois River. By 1994, zebra mussels had indeed colonized the length of the Mississippi River and many major tributaries (Ludyanskiy *et al.* 1993, Mackie & Schloesser 1996).

The remarkably rapid spread of zebra mussels through the Great Lakes and connected waterways prompted warnings that the colonization by zebra mussels of all suitable freshwater habitats across North America

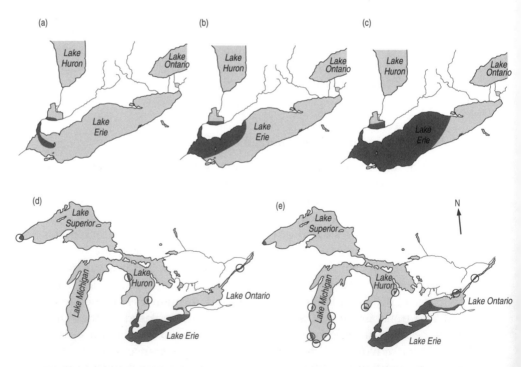

Fig. 13.4 The distribution of zebra mussels in the Great Lakes at the end of (a) 1986, (b) 1987, (c) 1988, (d) 1989, and (e) 1990. Dark areas indicate main zebra mussel range. Open circles indicate establishment of isolated populations. (After Griffiths *et al.* 1991 with permission of the National Research Council of Canada.)

would have occurred by the year 2000 (Ludyanskiy *et al.* 1993). The initial phase of zebra mussel invasion was, however, primarily an invasion of waterways directly connected to the Great Lakes system. Dispersal of aquatic animals through connected aquatic systems is easier than dispersal into isolated systems which involves at least some overland travel. Successful colonization following overland travel is less likely due to the potential for stress following lengthy periods out of water, and the low numbers of individuals that are likely to be transported during any particular dispersal event.

Kraft and Johnson (2000) hypothesized that the invasion of zebra mussels into more isolated waterbodies across North America would occur at a slower rate than initial predictions based on the rate of zebra mussel spread through connected waterways. Numerous isolated lakes of varying sizes are dotted across the landscape surrounding the Great Lakes system. Physical and chemical conditions are suitable for the survival of zebra mussels in most of them. However, if zebra mussels are to colonize, then some vector is required to transport the mussels overland. Kraft and Johnson (2000) sampled 140 isolated lakes in 1995, 1996 and 1997 in the region surrounding Lake Michigan. To determine whether zebra mussels had colonized, plankton samples were examined for the presence of the planktonic veliger larvae. One or two vertical plankton tows from depths of 4–6 m for a total sample volume of 800 L were collected from each lake. Sampling was conducted in spring, summer or autumn.

Over a 3-year period (1995–97), Kraft and Johnson (2000) detected zebra mussels in only 19% of the 140 lakes surveyed. The low proportion of lakes colonized occurred despite the relatively close proximity of the Great Lakes, a system largely colonized by zebra mussels by 1995. Marked regional differences in the proportion of lakes colonized by zebra mussels were observed. Zebra mussels were detected in seven of the 28 Indiana lakes surveyed, 15 of the 49 Michigan lakes surveyed, but only in five of the 63 Wisconsin–Illinois lakes surveyed (Fig. 13.5). No obvious environmental variables could account for the distribution pattern, other than a tendency for larger lakes to be infested.

The primary means by which zebra mussels are introduced to isolated waterbodies is by the overland transport of recreational trailer boats. The haphazard pattern of infestation observed by Kraft and Johnson (2000) may reflect the risky nature of overland travel for zebra mussels, and the highly variable patterns of boat use from region to region. What the study does show is that zebra mussel infestation into isolated waters occurs at a slower rate than that which occurs through connected waters. Whilst little can be done to limit the downstream spread of mussels through connected systems, education of boat users could potentially slow the spread of the mussels into isolated systems. Simple actions by boat users

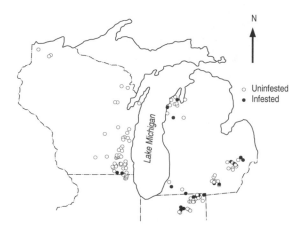

Fig. 13.5 Lakes surveyed as part of study by Kraft and Johnson (2000) across the lower peninsula of Michigan, Wisconsin, northern Indiana and northern Illinois. Lakes infested with zebra mussels are shown with solid circles, and lakes that were uninfested as of 1997 are shown with open circles. (After Kraft & Johnson 2000 with permission of the National Research Council of Canada.)

which could slow the rate of new infestations include draining of bilge water, removal of weed, and checking for and removing attached mussels on hulls and equipment prior to the transfer of boats from one waterbody to another.

13.3.5   What do zebra mussels do to freshwater ecosystems?

Determining the impact of zebra mussels on freshwater ecosystems is not simply a matter of searching for changes in randomly chosen measures of water quality or community structure that occur after zebra mussels invade. Given the extensive knowledge of zebra mussel ecology, it is possible to generate hypotheses predicting features of ecosystems that are likely to be altered by zebra mussels. We can begin to develop a conceptual model of likely impacts of zebra mussels by considering their position in freshwater food webs.

Zebra mussels are filter feeders that remove fine particles from the water column. Their primary food sources are phytoplankton, bacteria and small zooplankton. If zebra mussels are present in sufficient numbers, then it might be expected that the densities of their preferred food will decline. Phytoplankton is a primary food for a wide variety of other freshwater consumers, particularly large filter-feeding zooplankton. Declines in phytoplankton could result in reduced zooplankton densities, which may in turn impact on zooplanktivorous fish densities. Increased water clarity, due to the removal of fine particles from the water column, may allow increased light penetration and enhanced growth of aquatic plants. Increased availability of nutrients formerly taken up by phyto-plankton may further enhance aquatic plant growth, or alternatively enhance the growth rates of inedible phytoplankton. Zebra mussels also alter the nature of the benthic substrates they colonize, a result which can have implications for other benthic organisms.

Having a set of hypotheses that describe the likely impacts of zebra mussels allows targeted investigation of responses of key parameters to zebra mussel invasion. Two broad research strategies may be employed to examine the impact of mussels. We can conduct before and after comparisons of key variables in habitats in which a zebra mussel invasion has occurred. As with any descriptive study, the results must be inter-preted with caution given that other unknown factors may also be influ-encing the variables of interest. Alternatively, we can conduct more precise studies of zebra mussel impacts in the controlled environment of a laboratory or field enclosure. Again, such studies must also be interpreted with caution given that it requires extrapolation of impacts from a small simple system to larger more complex systems.

13.3.6 Comparing an ecosystem before and after a zebra mussel invasion

The North American invasion of zebra mussels occurred through some of the most intensively utilized and studied waterbodies in the world. Extensive long-term data sets covering a variety of water quality param-eters exist for many North American waters. Strayer *et al.* (1996, 1999) examined the impact of zebra mussels on the downstream reaches of the Hudson River by comparing records of key variables in the years before and after invasion by zebra mussels. An isolated population of zebra mussels was first observed in the Hudson River in May 1991 (Strayer *et al.* 1996). The zebra mussel invasion then followed the typical pattern with rapid downstream spread, and the appearance of isolated populations upstream of the initial infestation over the 2 following years. Strayer *et al.* (1999) compared information on key variables likely to be impacted by zebra mussels before and after the invasion. They also examined changes in other key environmental variables such as discharge and temperature to determine whether other factors might account for any changes that occurred in the river system over the study period.

Marked changes in the structure of the Hudson River ecosystem consist-ent with the predicted impact of zebra mussels did occur after 1992. Sub-stantial drops in chlorophyll a concentration (a measure of phytoplankton abundance), and declines in the densities of zooplankton and other species of bivalve were observed from 1993 onwards (Fig. 13.6a,b,c,d). Substantial increases in concentrations of nutrients such as nitrogen and phosphorus, and increased light penetration also occurred (Fig. 13.6e,f,g).

199

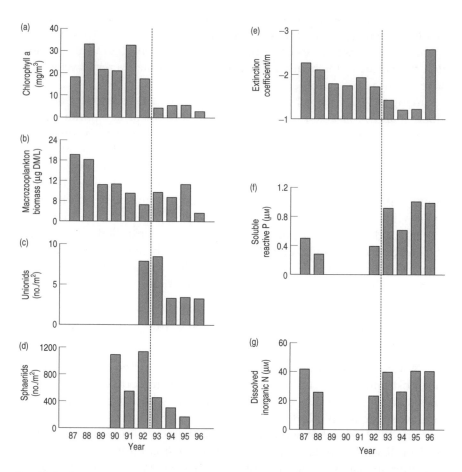

Fig. 13.6 Changes in (a) chlorophyll a concentrations, (b) macrozooplankton biomass, (c) density of unionid mussels, (d) density of sphaerid mussels, (e) extinction coefficient of light, (f) soluble reactive phosphorous concentration, and (g) dissolved inorganic nitrogen concentration after the arrival of zebra mussels in the Hudson River. Data are annual means from Kingston, New York. (After Strayer *et al.* 1999. Copyright, American Institute of Biological Sciences.)

The observed changes to the Hudson River ecosystem were not correlated with other environmental factors such as temperature or discharge. Similar changes in phytoplankton and zooplankton have also been observed in other lake and river habitats following zebra mussel invasions (see Strayer *et al.* 1999). Based on these various lines of evidence, Strayer *et al.* (1999) concluded that zebra mussels were the most likely cause of the observed changes to the Hudson River ecosystem over the period of study.

13.3.7 Experimental studies of zebra mussel impact

The changes in the Hudson River and other ecosystems before and after the invasion of zebra mussels are consistent with the predicted impact of zebra mussels based on their position within freshwater food webs. The observation of the same general pattern in other waterways invaded by zebra mussels certainly strengthens the inference that zebra mussels have significant impacts on aquatic ecosystems. However, as with any descriptive study, the possibility remains that an alternative unmeasured factor is the mechanism behind the observed changes.

To test specifically the hypothesis that zebra mussels reduce phytoplankton and zooplankton densities, an experiment manipulating zebra mussel densities under controlled conditions is required. Jack and Thorp (2000) enclosed three densities of mussel in floating enclosures (0, 1000 and 2500 mussels per enclosure, five replicates of each treatment). The enclosures were designed to permit some exchange of water with the surrounding Ohio River. Samples of zooplankton, phytoplankton and bacteria were collected over 6 consecutive days. Phytoplankton densities were estimated by direct counts. Growth rates of zooplankton populations were estimated by comparing initial and final densities. The results of their experiment suggested that the impact of the zebra mussels on the plankton communities within the enclosures was dramatic. Total counts of phytoplankton cell densities declined with increasing zebra mussel density, with cyanobacterial abundance showing the greatest reduction (Fig. 13.7a). Population growth rates of all crustacean zooplankton taxa were negative in enclosures containing zebra mussels (Fig. 13.7b).

13.4 Science and the management of invasive species

The invasion of ecosystems by exotic species is a global phenomenon, and the subject of many scientific investigations. The example of the zebra mussel invasion outlined above details some of the questions and approaches that an ecologist might use to understand such a problem. It is important to recognize the role and limitations of science in such investigations. Science can only provide us with information on the extent and impact of an invasion. It cannot tell us whether such an invasion is good or bad. Such a decision is a value judgement that is hopefully based on the high-quality information that a detailed scientific study can provide. Perspectives of whether a zebra mussel invasion is a good or bad event may vary. Those interested in maintaining biodiversity and intact native plant and animal communities will almost unanimously see the zebra mussel invasion in a negative light. The spread of zebra

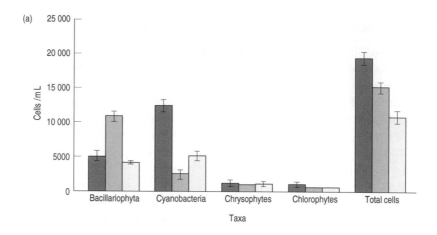

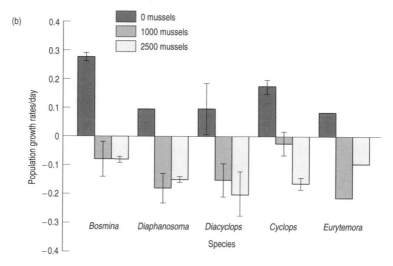

Fig. 13.7 Mean (± 1 standard error) phytoplankton cell density (major algal groups) in three mussel treatments on day 6 of sampling (a), and (b) mean population growth rate of crustacean zooplankton between day 1 and day 6 in zebra mussel enclosure experiments. See text for further details. (After Jack & Thorp 2000.)

mussels and other invasive species such as *Gambusia* is slowly eroding the unique features of different animal and plant communities across the planet. Alternatively, zebra mussels can improve water quality under certain circumstances, a property that some water managers might see as beneficial.

Scientists have a duty to provide objective and unbiased information to the best of their professional abilities. Some would argue that the role of the scientist is then finished. It is for professional managers to decide how to handle problems such as the zebra mussel invasion. However, whilst a scientific methodology does impose strict and often frustrating limitations on what we can say about our area of study, there would be few scientists who did not hold strong opinions on the importance and relevance of their own area of study. In many cases, those opinions are value judgements, and scientists should have a sufficient understanding of the scientific method to understand the difference between a scientific statement and a value judgement. Scientists should not feel compelled to state how their work should be used. However, the opinions of scientists who do express such value judgements are the opinions of those who have a direct experience of the environments in which they work. Such 'hands on' experience can prove invaluable.

13.5 Further reading

Environment Protection Service (1981) *The Presence and Implications of Foreign Organisms in Ship Ballast Water Discharged into the Great Lakes*, Vols 1 and 2. Environment Canada.

Griffiths R.W., Schloesser D.W., Leach J.H. & Kovalak W.P. (1991) Distribution and dispersal of the zebra mussel (*Dreissena polymorpha*) in the Great Lakes region. *Canadian Journal of Fisheries and Aquatic Sciences* **48**, 1381–1388.

Hebert P.D.N., Muncaster B.W. & Mackie G.L. (1989) Ecological and genetic studies on *Dreissena polymorpha* (Pallas): a new mollusc in the Great Lakes. *Canadian Journal of Fisheries and Aquatic Sciences* **46**, 1587–1591.

Jack J.D. & Thorp J.H. (2000) Effects of the benthic suspension feeder *Dreissena polymorpha* on zooplankton in a large river. *Freshwater Biology* **44**, 569–579.

Kraft C.E. & Johnson L.E. (2000) Regional differences in rates and patterns of North American inland invasions by zebra mussels (*Dreissena polymorpha*). *Canadian Journal of Fisheries and Aquatic Sciences* **57**, 993–1001.

Ludyanskiy M.L., Mcdonald D. & MacNeill D. (1993) Impact of the zebra mussel, a bivalve invader. *BioScience* **43**, 533–544.

Mackie G.L. (1991) Biology of exotic zebra mussel, *Dreissena polymorpha*, in relation to native bivalves and its potential impact in Lake St. Clair. *Hydrobiologia* **219**, 231–268.

Mackie G.L. & Schloesser D.W. (1986) Comparative biology of zebra mussels in Europe and North America: an overview. *American Zoologist* **36**, 244–258.

Strayer D.L. (1991) Projected distribution of the zebra mussel, mussel, *Dreissena polymorpha*, in North America. *Canadian Journal of Fisheries and Aquatic Sciences* **48**, 1389–1395.

Strayer D.L., Powell J., Ambrose P., Smith L.C., Pace M.L. & Fisher D.T. (1996) Arrival, spread, and early dynamics of a zebra mussel (*Dreissena polymorpha*) population in the Hudson River estuary. *Canadian Journal of Fisheries and Aquatic Sciences* **53**, 1143–1149.

Strayer D.L., Caraco N.F., Cole J.J., Findlay S. & Pace M.L. (1999) Transformation of freshwater ecosystem by bivalves. *BioScience* **49**, 19–27.

Concluding remarks

Freshwater ecology is a diverse subject. The research described in this text deals with only a small proportion of possible topics that falls within the realm of freshwater ecology. Yet even in this introductory text, we have dealt with research topics that require specialist skills in the areas of sampling of fish and invertebrate communities, analysis of water chemistry, and the identification of freshwater invertebrates and algae. As we move from one project to another, some of the technical skills acquired during previous research will be used again. Other new skills that are specific to the latest work will have to be learnt. For example, we have hardly touched upon the growing fields of aquatic microbiology or remote sensing and landscape ecology. The constant need to refine old skills and acquire new ones contributes significantly to the challenge and enjoyment that comes from doing scientific research. Further pleasure comes from collaborating with a diverse range of other scientists all over the world with additional specialist skills that we ourselves will never have the time to master. Collaboration permits us to broaden the scope and range of our own research without having to learn the detailed intricacy of new techniques. It also opens our minds to cultural differences in the way that scientists view their work and communicate their findings to managers, politicians and the public.

Another level of satisfaction is related to developing and refining skills that are common to all freshwater ecological research and ecological research in general. At the core of doing ecological research lies the skills associated with using the 'tools' of science. These include an intimate knowledge of how the scientific method works, an understanding of how scale influences our ability to detect pattern, the field-related skills of careful observation and natural history, an ability to collect data with consistency and rigour, and finally an ability to analyse, interpret and communicate our results effectively in scientific reports and publications. Scientists will use and refine their skills in the use of these 'tools' over the course of each project and over their entire research careers. As we conduct our own research and read the research of others, it is crucial that we recognize these skills so that we can develop them and improve our abilities as researchers.

The need for talented and knowledgeable freshwater ecological scientists is huge. As human populations grow, we are placing ever greater demands on our environment. We need researchers interested in fundamental ecological questions who are able to tease apart the workings of natural environments so that we can ensure that the systems that support our environment either remain in place or can be repaired. We also need researchers who are interested in applied ecological questions. Such scientists can design research that allows us to identify the cause of specific problems and the development of possible solutions. However, it is not enough for only scientists to understand the process of science. Understanding the strengths and limitations of ecological science is crucial for anyone who either works in, or is interested in, the area of environmental management. Understanding how science works is central to interpreting, critiquing and responding effectively in a flexible manner to the challenging and exciting demands of new questions and environmental problems that are always there to solve. What kind of scientist are you?

Index

Page numbers in *italics* refer to figures, those in **bold** refer to tables and boxes

Index

Index

Index

Index

Index